U0157772

计算机专业英语

主　编　郭　欣　杜　悦
副主编　张恒达　朱黎黎
主　审　陈永庆　郑志刚　李桂杰　王　锐

北京理工大学出版社
BEIJING INSTITUTE OF TECHNOLOGY PRESS

内 容 简 介

本书致力于为读者提供一套全面、系统且易于理解的计算机专业英语知识体系，充分借鉴当今国外和国内信息技术发展现状并结合计算机专业英语教学特点，在选材上从 IT 行业发展比较新的领域进行选择，凸显专业特色，并满足院校学生计算机相关大赛与认证考试需求。

本书内容涵盖了计算机专业的各个方面，包括计算机硬件、计算机软件、计算机语言、计算机安全、互联网、大数据、AI、物联网、区块链、职场与赛证等。在学习模块上基于计算机行业职业情境设计口语对话，选择 IT 行业新兴和热门领域科普文章作为单元阅读主题文章，并搭配相应内容图片音视频等内容，多元化呈现阅读文章内容。选择英语学习中所需的重要语法项目，并设置相应练习。以英语应用文为主，提供写作格式、范例、习题等。

本书适用于计算机专业的学生以及从事计算机相关领域工作的人员。

版权专有 侵权必究

图书在版编目（C I P）数据

计算机专业英语 / 郭欣，杜悦主编. -- 北京：北京理工大学出版社，2024.3

ISBN 978 - 7 - 5763 - 3325 - 1

Ⅰ. ①计⋯ Ⅱ. ①郭⋯ ②杜⋯ Ⅲ. ①电子计算机 - 英语 - 教材 Ⅳ. ①TP3

中国国家版本馆 CIP 数据核字（2024）第 020224 号

责任编辑：龙 微　　**文案编辑：**龙 微
责任校对：刘亚男　　**责任印制：**施胜娟

出版发行 / 北京理工大学出版社有限责任公司
社　　址 / 北京市丰台区四合庄路 6 号
邮　　编 / 100070
电　　话 / （010）68914026（教材售后服务热线）
　　　　　　（010）68944437（课件资源服务热线）
网　　址 / http：//www.bitpress.com.cn

版印次 / 2024 年 3 月第 1 版第 1 次印刷
印　　刷 / 三河市天利华印刷装订有限公司
开　　本 / 787 mm×1092 mm　1/16
印　　张 / 17
字　　数 / 378 千字
定　　价 / 85.00 元

图书出现印装质量问题，请拨打售后服务热线，负责调换

前言

在 21 世纪的信息时代，计算机科学技术在各个领域的应用日新月异，引领着全球技术革新的潮流。计算机专业英语，作为连接国际先进技术与语言交流的桥梁，其重要性不言而喻。为了满足广大计算机专业学生和从业者对国际前沿技术的求知需求，我们编写了这本《计算机专业英语》教材。

一、教材背景与目的

随着全球化的深入发展，计算机技术的国际交流与合作日益频繁。掌握专业英语，对于计算机领域的学者、工程师和研究人员来说，是获取国际最新研究成果、参与国际学术交流的必备技能。本教材正是为了帮助读者提高计算机专业英语水平，使读者具备国际视野、跨文化交流能力，掌握最新的 IT 行业知识，并满足读者各种大赛与认证考试的需求。

二、内容组织与特点

本教材充分借鉴当今国外和国内信息技术发展现状并结合计算机专业英语教学特点，以培养读者可持续发展的职业核心竞争力为目标，在内容选材上从 IT 行业发展比较新的领域进行选择，使得教材凸显专业特色，并满足读者大赛与认证考试需求。

教材包含十个单元，主题包括：计算机硬件、计算机软件、计算机语言、计算机安全、互联网、大数据、AI、物联网、区块链、职场与赛证。

每单元设置四个模块，结构如下：

Section Ⅰ. 听说部分

基于计算机行业职业情境设计口语对话。

Section Ⅱ. 阅读部分

选择 IT 行业新兴和热门领域科普文章作为单元阅读主题文章三篇，每篇主题文章搭配相应内容图片音视频等内容，多元化呈现阅读文章内容，并搭配文章翻译和练习题。

Section Ⅲ. 语法部分

选择英语学习中所需的重要语法项目，并设置相应练习。

Section Ⅳ. 写作部分

以英语应用文为主，提供写作格式、范例、习题等。

本教材的特点在于：

1. 内容全面：涵盖了计算机科学的各个方面，满足不同层次读者的需求。

2. 实用性强：注重培养读者的实际应用能力，通过丰富的实例和练习题提高读者的专业英语应用水平。

3. 结构清晰：章节设置合理，层次分明，方便读者系统地学习。

4. 难度适中：针对不同水平的读者，设置了不同难度的内容，便于读者逐步提高。

三、适用对象与学习建议

本教材适用于计算机及相关专业的大学生以及对计算机专业英语感兴趣的读者。对于初学者，建议从基础篇开始学习，逐步提高；对于有一定基础的读者，可以根据自己的兴趣和需求选择相应的章节进行深入学习。同时，结合实际应用场景进行实践练习，将有助于巩固所学知识，提高专业英语的实际应用能力。

四、期望效果与致谢

通过学习本教材，读者将能够熟练掌握计算机专业英语的基本词汇和表达方式，提高对英文技术文献的阅读和理解能力；同时，读者能够在国际学术交流和技术合作中自如运用计算机专业英语，成为具有国际视野和跨文化交流能力的优秀计算机人才。

本教材由渤海船舶职业学院长期奋战在教学一线的骨干教师编写完成。其中杜悦老师负责编写第一、二单元（包括翻译及答案），郭欣老师负责编写第三、四、九、十单元（包括翻译及答案），朱黎黎老师负责编写第五、六单元（包括翻译及答案），张衡达老师负责编写第七、八单元（包括翻译及答案）。

在教材编写过程中，我们得到了众多专家、学者和一线教师的支持与帮助，在此向他们表示衷心的感谢！同时感谢为本书提供素材和技术支持的相关单位和个人。此外，我们还要感谢为本书付出辛勤劳动的编辑和工作人员，正是有了你们的努力和支持，才使本书得以顺利出版。

由于编者水平有限且时间仓促，疏漏之处在所难免，恳请广大同仁批评斧正。

编　者

2024. 1. 16

目录

Unit 1

Computer Hardware

Unit Goals

After learning this unit, you will be able to:

- talk about computer hardware
- understand the passages and grasp the key words and expressions
- review the grammar of the part of speech
- write a telephone message

Section I Listening & Speaking

Part A Words and Expressions

RAM（Random Access Memory） 随机存取存储器	imagination 想象力；想象
GB 千兆字节（计算机存储单位）	personality 性格；个性；气质
Mbps 兆比特每秒（传输速率单位）	possess 有；拥有；具有（特质）
cursor 光标，游标	applicant 申请人
jerk 急拉；猛推；猝然一动	AutoCAD 三维辅助设计软件
programming 编程	certificate 证明；合格证书；文凭
quality 质量；品质	original 起初的；原来的；原件

Part B Conversations

Read the conversations carefully and then complete the communicative tasks.

Conversation 1

A： I thought computers were supposed to make your life more convenient. But it seems I spend half my time waiting for it to do something.

B： Why is it so slow? How much **RAM** do you have?

A： I think I have 4 **GB** RAM.

B： That's your problem. If you want Windows 10 to run smoothly, you need at least 16 GB RAM.

A： I have 16 GB RAM now and most things work better, but my Internet connection is still really slow. Is there anything that I can do?

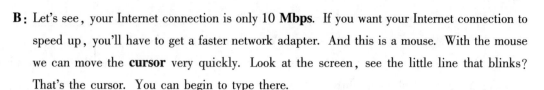

B： Let's see, your Internet connection is only 10 **Mbps**. If you want your Internet connection to speed up, you'll have to get a faster network adapter. And this is a mouse. With the mouse we can move the **cursor** very quickly. Look at the screen, see the little line that blinks? That's the cursor. You can begin to type there.

A： But I don't want to type there.

B： Then move the cursor. Watch, I'll hold the mouse in my hand. Now I'll move it across the table. See the cursor move?

A： I can't get my mouse work properly. First it moves slowly and then, all of a sudden, it **jerks** all the way across the screen. What can I do?

B： Here, let me see. (Open the mouse and remove the ball.) Do you see these contacts here?

A： Yes, they look pretty dirty.

B： That's right. Let me clean those for you, and it'll be work in no time.

 Key Words

RAM（Random Access Memory）　随机存取存储器
GB 千兆字节（计算机存储单位）
Mbps　兆比特每秒（传输速率单位）
cursor　光标，游标
jerk　急拉；猛推；猝然一动

Task 1　Act out the conversation with your partner based on the following clues.

A： I thought computers _____①_____. But it seems I spend half my time waiting for it to do something.

B： Why is it so slow? _____②_____?

A： I think I have 4 GB RAM.

B： That's your problem. If you want Windows 10 to run smoothly, you need at least 16 GB RAM.

A： I have 16 GB RAM now and most things work better, but my Internet connection is still really slow. _____③_____?

B： Let's see, your Internet connection is only 10 Mbps. If you want your Internet connection to speed up, you'll have to get a faster network adapter. And this is a mouse, With the mouse we can move the cursor very quickly. Look at the screen, see the little line that blinks? That's the cursor. You can begin to type there.

A： But I don't want to type there.

B： Then move the cursor. Watch, _____④_____. Now I'll move it across the table. See the cursor move?

A：I can't get my mouse work properly. First it moves slowly and then, _____⑤_____ , it jerks all the way across the screen. What can I do?

B：Here, let me see. (Open the mouse and remove the ball). Do you see these contacts here?

A：Yes, they look pretty dirty.

B：That's right. Let me clean those for you, and It'll be work in no time.

Conversation 2

A：Do you have any work experience in IT companies?

B：Yes, I used to have a part-time job in an IT company.

A：What's your main duty there?

B：Computer **programming**.

A：What **qualities** do you think a computer programmer should have?

B：In my opinion, a computer programmer should have teamwork spirit and an active **imagination**. He should be talented in creation and sensitive to the changes in IT market.

A：How did your previous employers treat you?

B：They treated me very well. We cooperated harmoniously and respected each other.

A：What have you learned from the part-time job you have had?

B：I have learned how to behave myself as a qualified employee and some skills of how to get along with people with different **personalities**.

 Key Words

programming 编程
quality 质量；品质
imagination 想象力；想象
personality 性格；人格；气质

Task 2 Act out the conversation with your partner based on the following clues.

A：Do you have any work experience in IT companies?

B：Yes, _____①_____ .

A：What's your main duty there?

B：_____②_____ .

A：What qualities do you think a computer programmer should have?

B：_____③_____ , a computer programmer should have teamwork spirit and an active imagination. He should be talented in creation and sensitive to the changes in IT market.

A：How did your previous employers treat you?

B：_____④_____ . We cooperated harmoniously and respected each other.

A： _____⑤_____ ?

B： I have learned how to behave myself as a qualified employee and some skills of how to get along with people with different personalities.

Conversation 3

A： What qualifications do you **possess** for the present job?

B： I have four years' study in the Department of Software and it has given me a solid theory foundation. Moreover, I have worked part-time at an IT company for 2 years and got a lot of practical experience.

A： We have several **applicants** for this position. Why do you think you are the person we should choose?

B： I have the qualification and experience that you require. For example, I had two years' experience in programming and got leadership experience while serving the college student union as president.

A： What kind of software can you use?

B： I am adept at DOS, Windows, and **AutoCAD**. I am also pretty familiar with Fortran and C language.

A： Do you have any **certificates** on computer?

B： Yes, I have NCRE certificate, rank 1.

A： Do you take the **original** certificate with you?

B： Yes. Here it is.

Key Words

possess 有；拥有；具有（特质）

applicant 申请人

AutoCAD 三维辅助设计软件

certificate 证明；合格证书；文凭

original 起初的；原件；正本

Task 3 Act out the conversation with your partner based on the following clues.

A： What qualifications do you possess for the present job?

B： _____①_____ and it has given me a solid theory foundation. Moreover, I have worked part-time at an IT company for 2 years and got a lot of practical experience.

A： _____②_____ . Why do you think you are the person we should choose?

B： _____③_____ . For example, I had two years' experience in programming and got leadership experience while serving the college student union as president.

A：_____④_____?

B：I am adept at DOS, Windows, and AutoCAD. I am also pretty familiar with Fortran and C language.

A：_____⑤_____?

B：Yes, I have NCRE certificate, rank 1.

A：Do you take the original certificate with you?

B：Yes. Here it is.

Part C Passages

听力材料

Listen to the following passages carefully and fill in the blanks with the information you've heard.

Passage 1

Hardware is a computer _____ that users can touch directly. There are four main hardware devices in the computer _____, which can be divided into host devices and _____ devices. Host devices are generally inside the main box, including: central processing unit (CPU), main board, _____, display card, sound card, hard disk, optical disk _____ and other components.

Passage 2

The _____ of assembling a computer is not difficult. All parts in the computer are _____ and manufactured according to _____ protocols. Now the assembly process is much more _____. Let's briefly _____ the assembly process of the computer.

Passage 3

Open the tenon of the _____ slot on the motherboard. Look at each memory module and place it on the slot so that the notch at the _____ of the memory module matches and aligns with the bump on the memory slot. Finally, _____ the memory module down with a little _____ until it snows into place, then the tenon closes by _____.

Passage 4

Next, we will put the _____ with CPU, memory and solid-state SSD hard disk _____ into the case. _____ sure that the holes on the mother board are aligned with the studs you installed, and the ports are aligned with the holes on the I/O shielding bezel. _____ the mother board is in place, screw the screws into the nut posts and _____ the mother board in place.

Passage 5

After completing all _____ and careful inspection, plug in and _____ the monitor, keyboard and mouse, and then plug in the power supply.

_____ the power button on the monitor, then turn on the power switch (located on the back of the power supply), and then press the power button of the computer. If everything is _____, the computer should _____ on and run POST.

Section Ⅱ Reading

Passage 1

Computer Hardware

Hardware is a computer **component** that users can touch directly. There are four main hardware **devices** in the computer system, which can be divided into host devices and **external** devices. [①] They are:

Host devices: They are generally inside the main box, including central processing unit (CPU), main board, memory, display card, sound card, hard disk, **optical** disk drive and other components.

External devices: Relative to the **case**, it is located outside the case, including various **input/output (I/O)** devices, external **storage** devices, etc.

1. CPU

The CPU composed of transistors is the **core** for processing data and executing programs. Its full English name is Central Processing Unit (CPU). First, the internal structure of CPU can be divided into three parts: control unit, logic operation unit and storage unit (including internal bus and buffer). The working principle of CPU is like the processing of products in a factory: raw materials (program instructions) entering the factory is dispatched and distributed by the material distribution department (control unit), sent to the production line (logic operation unit), and finished products (processed data) are produced, stored in the warehouse (storage unit), and finally sold in the market (for used by the application program). The dominant frequency is an important indicator to measure the CPU's operation speed. The higher the dominant frequency value is, the faster the CPU works. [②]

2. Main board

The main board, also known as the mother board or the system board, is the carrier of various components of the computer. With the main board, the CPU can issue commands, and various devices can communicate in this way. [③] It is closely connected with the computer to form an **organic** whole, through which the performance of various accessories can be played.

The mother board adopts an open structure. There are 6 – 15 **expansion slots** on the mother board, which are used to plug control cards (adapters) of PC **peripherals**. By replacing these cards, the corresponding **subsystems** of the microcomputer can be partially upgraded, so that manufacturers and users have greater **flexibility** in configuring models. In a word, the mother board plays an important role in the whole microcomputer system. [④] It can be said that the type and grade of the mother board determines the type and grade of the entire microcomputer system, and the performance of the mother board affects the performance of the entire microcomputer system.

3. Memory

Memory is one of the most important components in a computer, and it is the bridge to communicate with CPU. All programs in the computer run in memory, so the performance of memory has a great impact on the computer. [⑤] Memory, also known as internal memory, is used to temporarily store computing data in the CPU and data exchanged with external memory such as hard disk. As long as the computer is running, the CPU will transfer the data to the memory for calculation. When the calculation is completed, the CPU will transfer the results. The size and stable operation of the memory also determine the performance of the computer.

4. Storage System

The storage system is used to store data. During the operation of the computer, memory is used to store the data in the current work task of the user. Because memory is only a temporary storage device, users must save relevant data files to the relevant storage device before ending program running or shutting down the computer.

The storage device used is determined by the amount of data, the speed of data retrieval or the speed of data **transmission**. At present, there are mainly mechanical hard disk, **solid state hard disk**, optical disk, U disk, flash memory card and other storage systems.

5. Monitor

A display is an output device used to display information in a computer. All monitors have a power switch and a set of control switches for adjusting screen brightness and **contrast**.

The monitor has a variety of models, **resolutions** and sizes. The larger the display, the larger the image. The resolution determines the **clarity** and **accuracy** of the image presentation.

6. Printer

Printer is one of the common output devices of computer, which is used to print the computer processing results on relevant media. Generally speaking, the impact printer and non-impact printer can be divided according to whether the printing element strikes the paper. According to the imaging technology used, there are **color separation ribbon, ink-jet, laser, thermal, electrostatic** and other printers.

At present, there are many types of printers. The type of printer depends on the needs of users. For example, format paper or checks may require dot matrix printers, while documents such as letters and budget reports may require laser printers.

New Words and Expressions

component [kəmˈpəʊnənt]　*n.* 组成部分；成分；部件

device [dɪˈvaɪs]　*n.* 装置；仪器；设备

external [ɪkˈstɜːnl]　*adj.* 外部的　*n.* 外部；外观

optical [ˈɒptɪkəl]　*v.* 转寄；发送

case [keɪs]　*n.* (车辆的)底盘；底座

input/output (I/O) [ˈɪnpʊt]/[ˈaʊtpʊt]　*n.* 输入/输出

storage [ˈstɔːrɪdʒ]　*n.* 存储；保管；存储能力

core [kɔː(r)]　*n.* 中心部分；核心

organic [ɔːˈɡænɪk]　*adj.* 有机的　*n.* 有机体

expansion [ɪkˈspænʃn]　*n.* 膨胀；扩张；扩展

slot [slɒt]　*n.* 插槽；狭槽

peripheral [pəˈrɪfərəl]　*n.* 外围设备；周边设备　*adj.* 与计算机相连的

subsystem [səbˈsɪstəm]　*n.* 子系统模块；子体系；子系统

flexibility [ˌfleksəˈbɪləti]　*n.* 灵活性；柔韧性；弹性

transmission [trænzˈmɪʃn]　*n.* 传输；传递；传送

solid state hard disk　固态硬盘

contrast [ˈkɒntrɑːst, kənˈtrɑːst]　*n.* 差异；对照　*v.* 对照

resolution [ˌrezəˈluːʃn]　*n.* 决议；解决

clarity [ˈklærəti]　*n.* 清晰；清楚；明确

accuracy [ˈækjərəsi]　*n.* 精确(程度)；准确(性)

color separation ribbon　分色色带

ink-jet [ɪŋk dʒet]　喷墨打印

laser [ˈleɪzə(r)]　*n.* 激光；激光器

thermal ［ˈθɜːml］ *adj.* 热的；保暖的；温暖的

electrostatic ［ˌɪlektrəʊˈstætɪk］ *adj.* 静电的

Notes

1. "There are four main hardware devices in the computer system, which can be divided into host devices and external devices."

   Analysis："which can be divided into host devices and external devices" 是 which 引导的非限制性定语从句，起到修饰主句的作用。

   Translation：计算机系统中有 4 种主要的硬件设备，可以将其分为主机设备和外部设备两类。

2. "The higher the dominant frequency value is, the faster the CPU works."

   Analysis：The + 比较级，the + 比较级，表示"越…越…"。

   Translation：主频数值越高，CPU 的工作速度越快。

3. "With the main board, the CPU can issue commands, and various devices can communicate in this way."

   Analysis："With the main board" with 复合结构表示"随着…"。

   Translation：有了主板，CPU 才可以发号施令，各种设备才能通过主板沟通。

4. "In a word, the mother board plays an important role in the whole microcomputer system."

   Analysis："plays an important role in…" 表示"在…中扮演着举足轻重的角色"。

   Translation：总之，主板在整个微机系统中扮演着举足轻重的角色。

5. "All programs in the computer run in memory, so the performance of memory has a great impact on the computer."

   Analysis："have \ has a great impact on…" 表示"对…有重要的影响"。

   Translation：计算机中所有程序的运行都是在内存中进行的，因此内存的性能对计算机的影响非常大。

Exercises

Ⅰ. Answer the following questions according to the text.

1. How many main hardware devices are there in the computer system?

   _____

   _____

2. What are host devices of a computer?

   _____

   _____

3. What are external devices of a computer?

4. Why do people like larger display?

5. How do users choose the type of printer?

II. Fill in the blanks with words according to the meaning of the article by memory.

The CPU composed of transistors is the _____ for processing data and executing programs. Its full English name is _____ Processing Unit. (CPU) First, the internal structure of CPU can be divided into three parts: control unit, logic operation unit and _____ unit (including internal bus and buffer). The working _____ of CPU is like the processing of products in a factory: raw materials (program instructions), entering the factory is dispatched and distributed by the _____ distribution department (control unit), sent to the production line (logic operation unit), and finished products (processed data) are produced, stored in the warehouse (storage unit), and finally sold in the market (for used by the _____ program). The dominant frequency is an important _____ to measure the CPU's operation speed. The higher the dominant frequency value is, the _____ the CPU works.

III. Fill in the blanks with the words given below. Change the forms when necessary.

transfer	model	divide
use	bridge	common
direct	adopt	

1. The motherboard _____ an open structure.

2. It is the _____ to communicate with CPU.

3. Hardware is a computer component that users can touch.

4. When the calculation is completed, the CPU will _____ the results.

5. The internal structure of CPU can be _____ into three parts: control unit, logic operation unit and storage unit.

6. The monitor has a variety of _____, resolutions and sizes.

7. Printer is one of the _____ output devices of a computer.

8. Memory is _____ to store the data in the current work task of the user.

IV. Translate the following sentences into Chinese.

1. There are four main hardware devices in the computer system, which can be divided into host

devices and external devices.

_____

_____

2. With the main board, the CPU can issue commands, and various devices can communicate in this way.

_____

_____

3. The size and stable operation of the memory also determine the performance of the computer.

_____

_____

4. The larger the display, the larger the image.

_____

_____

5. Printer is one of the common output devices of computer, which is used to print the computer processing results on relevant media.

_____

_____

Passage 2

The Assembly rocess of the Computer（Ⅰ）

The process of assembling a computer is not difficult. All parts in the computer are produced and manufactured according to standard **protocols**. Now the **assembly** process is much more convenient. Let's briefly explain the assembly process of the computer.

Before starting to assemble the computer, you need to have all the parts and tools ready：

Cross head screwdriver

Ties for cable arrangement

Flashlight（if necessary）

Heat dissipation paste（if necessary）

Sharp nose pliers

Computer components

A clean and stable installation platform

Note：Before we touch the computer **accessories**, we need to release the static electricity on our body to avoid short **circuit** of the computer or components![1]

In order to facilitate operation, we generally **install** the CPU, **heat sink**, memory and M2SSD solid state drive components on the motherboard before putting the motherboard into the case. [2]

1. Install CPU

Whether it is an Intel or AMD CPU, first release the tension lever on the motherboard CPU socket so that the processor can be placed in the CPU slot (pay attention to the installation direction, the CPU gap should correspond to the protrusion of the socket). The CPU is correctly seated in the slot. Press the tension lever and the CPU is tightly fixed in the slot.

2. Install CPU Cooler

The bracket of the CPU cooler is installed on the motherboard (Intel and AMD are different), a layer of heat dissipation paste is applied evenly and thinly on the surface of the CPU, the CPU cooler is stably placed on the surface of the CPU, the CPU cooler is fixed with the bracket, and the bottom surface of the CPU cooler is tightly combined with the surface of the CPU.

After the CPU cooler is installed, plug the radiator fan connector into the socket marked "CPU_FAN" on the motherboard.

3. Install Memory

Open the **tenon** of the memory slot on the motherboard. Check each memory module and place it on the slot so that the notch at the bottom of the memory module matches and aligns with the bump on the memory slot. Finally, press the memory module down with a little force until it snaps into place, so that the tenon closes by itself. [3]

4. Install M2 SSD Solid State Drive

Remove the **screws** of the M. 2 slot and slide into the SSD at a certain tilt angle. Make sure the notch is aligned with the slot, similar to RAM installation. ④ Slowly press the SSD flat and fix the mounting screws. Sometimes we have to install a separate radiator for M2 SSD SSDs.

New Words and Expressions

protocol ['prəʊtəkɒl]   *n.* 协议

assembly [ə'sembli]   *n.* 装配

screwdriver ['skruːdraɪvə(r)]   *n.* 改锥；螺丝刀

dissipation [ˌdɪsɪ'peɪʃn]   *n.* 消散；驱散

accessory [ək'sesəri]   *n.* 附件；配件   *adj.* 辅助的

circuit ['sɜːkɪt]   *n.* 环行；电路；线路

install [ɪn'stɔːl]   *v.* 安装；设置；建立(程序)

heat sink   散热片(器)

cooler ['kuːlə(r)]   *n.* 冷却器

socket ['sɒkɪt]   *n.* (电源)插座；(电器上的)插口

tenon ['tenən]   *n.* 凸榫；榫舌

screw [skruː]   *n.* 螺丝钉

align [ə'laɪn]   *v.* 排列；校准；排整齐

Notes

1. "Before we touch the computer accessories, we need to release the static electricity on our

body to avoid short circuit of the computer or components!"

Translation：在接触计算机配件之前，我们需要释放掉身上的静电，避免电脑或部件短路！

2. "In order to facilitate operation, we generally install the CPU, heat sink, memory and M2SSD solid state drive components on the motherboard before putting the motherboard into the case."

Translation：为了方便操作，我们一般在主板放入机箱之前，先将 CPU、散热器、内存和 M2SSD 固态硬盘部件安装到主板上。

3. "Finally, press the memory module down with a little force until it snaps into place, so that the tenon closes by itself."

Translation：最后，用一点力向下压内存条，直到它卡入到位，从而使卡榫自行关闭。

4. "Make sure the notch is aligned with the slot, similar to RAM installation."

Translation：确保槽口与插槽对齐，类似于 RAM 安装。

 Exercises

Answer the following questions according to the text.

1. Is the process of assembling a computer very difficult?

_____

_____

2. What kind of parts and tools do we need to have before starting to assemble the computer?

_____

_____

3. Why do we first release the tension lever on the motherboard CPU socket?

_____

_____

4. What do we do after the CPU radiator is installed?

_____

_____

5. What do we press the memory module down with a little for?

_____

_____

Passage 3

The Assembly Process of the Computer (Ⅱ)

1. Remove the Side Panel on the Case

First remove the side panel on the case. Generally speaking, there are thumb screws to fix the side panel, which is easy to disassemble. ①

2. Install Standoffs

Generally, nut posts for motherboard fixing will be provided in the case, and a suitable installation position will be found according to the actual situation of the motherboard. They may be marked on the case according to the size of the motherboard you choose. Generally speaking, the case has been pre-installed with studs. If the position is correct, skip this step. If the nut post is pre-installed in the wrong position, you can use pointed-nose pliers to take it out and reinstall it in the correct position. ②

3. I/O Shielding Baffle

The I/O shielding baffle is the I/O shielding cover in the area around the motherboard port, which is generally provided with the motherboard. Before putting the motherboard into the case, you need to install the I/O shielding baffle into the case, paying attention to the direction so that the motherboard port can pass through the hole. ③ In addition, there are some highly configured motherboards that are pre-installed with this I/O shielding baffle at the factory.

4. Put It in the Motherboard

Next, put the motherboard with CPU, memory and solid-state SSD hard disk installed into the case. Make sure that the holes on the motherboard are aligned with the studs you installed, and the ports are aligned with the holes on the I/O shielding baffle. Once the motherboard is in place, screw the screws into the nut posts and fix the motherboard in place.

5. Install the Power Supply

Thepower supply is usually installed on the top or bottom of the case, just screw it in place with four screws behind the case. Then, insert the 24-pin motherboard power connector and CPU power connector into the motherboard.

6. Install SATA Storage

We added M. 2 storage before. In order to expand the storage capacity, we can continue to install SATA storage, which can be a 2.5-inch SSD or a 3.5-inch hard disk drive. Connect the drive to the motherboard with a SATA data cable, and then connect the SATA power connector from the power supply to the drive. ④ Install the hard drive or SSD into the corresponding bracket, and then screw or snap it into place. Please note that the installation method and location of the bracket/ drive vary depending on the case model.

7. Install the Graphics Card

If you are using an Intel or AMD CPU with integrated graphics and do not plan to play large 3D games, then you may not need to install a graphics card. However, many AMD CPUs and Intel models do not have onboard graphics cards and require graphics cards to connect and output to the monitor.

To install the graphics card, you need to remove some slot covers on the back of the case so that the **HDMI** and DVI ports can be exposed and the monitor can be connected later.

Insert the graphics card into the PCIe X16 slot on the motherboard, and insert the PCIe power connector from the power supply into the card.

8. Case Panel Cable Connection

Before we turn on the computer, we need to connect the front panel audio cable, **USB** connector, power supply, reset, HDD activity indicator, etc. to the corresponding motherboard socket. At this time, you need to refer to the motherboard manual, because their location varies with the motherboard model.

9. Turn On the Computer

After completing all operations and careful inspection, plug in and connect the monitor, keyboard and mouse, and then plug in the power supply.

Click the power button on the monitor, then turn on the power switch (located on the back of the power supply), and then press the power button of the computer. If everything is normal, the computer should turn on and run **POST**.

10. Cable Arrangement

In order to make the interior of the case more beautiful and tidier, and to ensure better air circulation, we need to organize and bundle the cables and connecting wires inside the case.

11. Install the Operating System, Drivers and Updates.

Prepare a USB installation drive Windows 10 or Windows 11 (which can be made on Microsoft's official website).

Plug the USB drive into the new computer, turn on the power, and boot to the **operating system** installer, which will guide you through the process.

After you install the operating system, you need to install the drivers of various computer devices so that they can work properly.

Here, the assembly of a computer is over.

New Words and Expressions

case [keɪs]   *n.* 机箱

baffle [ˈbæfl]   *n.* 隔板；挡板

power supply   供电；电源

24-pin   24 孔电源接口标准

SATA   串行硬件驱动器接口

graphics [ˈɡræfɪks]   *n.* 图样；图像

HDMI   高清多媒体接口

USB   通用串行总线接口

POST   主机上电自检

operating system   操作系统

Notes

1. "Generally speaking, there are thumb screws to fix the side panel, which is easy to disassemble."

   Translation：一般来说，都有拇指螺钉将侧面板固定，这样便于拆卸。

2. "If the nut post is pre-installed in the wrong position, you can use pointed-nose pliers to take it out and reinstall it in the correct position."

   Translation：如果螺母柱预装在错误位置，可以使用尖嘴钳将其取出，重新按正确位置安装。

3. "Before putting the motherboard into the case, you need to install the I/O shielding baffle into the case, paying attention to the direction so that the motherboard port can pass through the hole."

   Translation：在将主板放入机箱中之前，您需要将 I/O 屏蔽挡板安装到机箱中，注意方向，以便主板端口可以穿过孔。

4. "Connect the drive to the motherboard with a SATA data cable, and then connect the SATA power connector from the power supply to the drive."

Translation：用SATA数据电缆将驱动器连接到主板，然后将SATA电源连接器从电源连接到驱动器。

5. "After you install the operating system, you need to install the drivers of various computer devices so that they can work properly. "

Translation：安装完操作系统后，需要再安装计算机各种设备的驱动程序，以便它们能正常工作。

 Exercises

Translate the following short passages into Chinese.

1. Next, put the motherboard with CPU, memory and solid-state SSD hard disk installed into the case. Make sure that the holes on the motherboard are aligned with the studs you installed, and the ports are aligned with the holes on the I/O shielding baffle. Once the motherboard is in place, screw the screws into the nut posts and fix the motherboard in place.

_____

_____

_____

2. After completing all operations and careful inspection, plug in and connect the monitor, keyboard and mouse, and then plug in the power supply. Click the power button on the monitor, then turn on the power switch (located on the back of the power supply), and then press the power button of the computer. If everything is normal, the computer should turn on and run POST.

_____

_____

_____

3. Prepare a USB installation drive Windows 10 or Windows 11 (which can be made on Microsoft's official website). Plug the USB drive into the new computer, turn on the power, and boot to the operating system installer, which will guide you through the process.

_____

_____

_____

Section III Grammar

Part of Speech（词性）

能够自由运用的最小语言单位叫词。根据词的形式、意义及其在句中的作用所作的分类叫词类（part of speech），又叫词性。英语的词通常分为十大类，即名词、冠词、代词、数词、形容词、副词、动词、介词、连词和感叹词。

一、名词

名词（noun）是表示人、事物、地点或抽象概念的名称，例如：Newton 牛顿 freedom 自由。

（一）名词可分为两大类：

1. 专有名词（proper noun）是特定的某人、地方或机构的名称。专有名词的第一个字母必须大写。

例如：China 中国 United Nations 联合国

2. 普通名词（common noun）是某类人、某种事物、某种物质或抽象概念的名称。

例如：teacher 教师 rice 大米

（二）普通名词又可分为可数名词（countable noun）与不可数名词（uncountable noun）两种。可数名词有单、复数之分。绝大多数名词的复数形式的构成是在单数名词的后面加-s 或-es。

例如：shop→shops 商店 bus→buses 公共汽车

英语中有一些名词的复数形式是不规则的。

例如：man→men 男人 tooth→teeth 牙齿

二、冠词

冠词（article）放在名词之前，帮助说明该名词所指的对象。冠词分为不定冠词（indefinite article）和定冠词（definite article）两种。

（一）不定冠词为 a/an，用在单数名词之前，表示某一类人或事物的"一个"。a 用在以辅音开头的名词之前，an 用在以元音开头的名词之前。

例如：a school 一所学校 an elephant 一头大象 an honest man 一个诚实的人

（二）定冠词只有一个，即 the，表示某一类人或事物中特定的一个或一些。可用于单数或复数名词前，也可用于不可数名词前。

例如：the girl 那个女孩 the Olympic Games 奥运会

三、代词

代词（pronoun）是用来指代人或事物的词。代词包括：

人称代词	例如：I, you, they, it 等
物主代词	例如：my, his, their, our, mine, hers 等

反身代词	例如：myself, yourself, itself, ourselves, oneself 等
相互代词	例如：each other, one another 等
指示代词	例如：this, that, these, those, such, same 等
疑问代词	例如：who, whom, whose, which, what 等
关系代词	例如：who, whom, whose, which, that 等
不定代词	例如：some, any, no, all, one, every, many, a little, someone, anything 等

四、数词

数词（numeral）是表示"数量"和"顺序"的词。分为基数词和序数词。

（一）基数词，例如：one（一），twenty（二十），thirty-five（三十五），one hundred and ninety-five（一百九十五）等。

（二）序数词，例如：first（第一），twentieth（第二十），fifty-first（第五十一）等。

五、形容词

形容词（adjective）是用来修饰名词，表示名词属性的词。例如：yellow（黄色的），wonderful（惊人的），strong（强大的）。形容词一般放在它所修饰的名词之前。

例如：busy streets（繁华的街道），public relations（公共关系），young men（年轻人）等。

形容词的比较等级可分为三种，即原级、比较级和最高级。

原级： （未变化的形容词原形）	比较级： （形容词 + 后缀-er 或 more + 形容词）	最高级： （形容词 + 后缀-est 或 most + 形容词）
great	greater	greatest
big	bigger	biggest
difficult	more difficult	most difficult

六、副词

副词主要用来修饰动词、形容词、副词或整个句子。可分为四种：

普通副词	例如：together（一起），well（好），carefully（仔细地）等
疑问副词	例如：when（何时），where（何地），how（如何），why（为什么）等
连接副词	例如：therefor（因此），then（然后），however（然而），otherwise（否则）等
关系副词	例如：where（何地），when（何时），why（为什么）等

副词比较等级的构成和形容词一样。

七、动词

动词（verb）是表示动作或状态的词，例如：sign（签字），support（支持），have（有），exist（存在）等。

动词根据意义和作用分为：实义动词、系动词、情态动词和助动词。

实义动词	例如：explain（解释），stay（停留），have（有）等
系动词	例如：be（是），seem（似乎），look（看起来），become（变成），get（变得）
情态动词	例如：can（能够），may（可以，也许），must（必须）
助动词	例如：shall，will，have，be，should，would，do 等

实义动词还可根据是否需要宾语分为及物动词和不及物动词。

及物动词（transitive verb）后面要跟宾语，意义才完整	例如：You must consider the matter carefully.
不及物动词（intransitive verb）本身意义完整，后面不需跟宾语	例如：The old man walked very slowly.

动词有四种基本形式，即动词原形、过去式、过去分词和现在分词。

原形	过去式	过去分词	现在分词
live	lived	lived	living
build	built	built	building
have	had	had	having

动词过去式和过去分词的构成有规则和不规则两种。规则动词（regular verb）的过去式和过去分词，在动词原形后面加-ed 或-d 构成。不规则动词（irregular verb）的过去式和过去分词的构成是不规则的，例如 eat，ate，eaten。这些动词数量虽不多，但都是比较常用的，必须熟记。

现在分词是在动词原形后面加-ing 构成。

八、介词

介词（preposition）又叫前置词，放在名词、代词或相当于名词的词前面，表示它后面的词与句子中其他成分之间的关系。

介词根据其构成，可分为：

简单介词	例如：in, at, for, since 等
复合介词	例如：into（进入），as for（至于），out of（出自）
二重介词	例如：until after（直至…之后），from among（从…当中）
短语介词	例如：according to（根据），because of（因为），in front of（在…之前）
分词介词	例如：regarding（关于），considering（考虑到），including（包括）

九、连词

连词（conjunction）是连接词、短语、从句或句子的词。连词是虚词，不能在句中单独作句子成分。

根据连词本身的含义及其所连接成分的性质，可分为并列连词和从属连词。

并列连词是连接并列关系的词、短语、从句或句子的连词	例如：and（和），or（或者、否则），but（但是），for（因为），not only...but also（不仅…而且），neither...nor（既不…也不）
从属连词是连接主从复合句的主句和从句的连词	例如：that，if（如果），whether（是否），when（当…时候），although（虽然），because（因为），so that（结果）

十、感叹词

感叹词（interjection）是表示喜怒哀乐等感情的词，例如：oh，well，why，hello 等。

Exercises

Choose the best answer.

1. – Would you like some drinks, boys?

   – Yes, _____ , please.

   A. some oranges   B. two boxes of chocolate

   C. some cakes   D. two boxes of cola

2. In China, September 10th is _____?

   A. Teacher Day   B. Teachers Day   C. Teacher's Day   D. Teachers' Day

3. This class _____ now. Miss Gao teaches them.

   A. are studying   B. is studying   C. be studying   D. studying

4. _____ live in Room 208.

   A. The Green   B. Green   C. The Greens   D. Greens

5. – Is this your shoe?

   – Yes, but where is _____?

   A. the other one   B. other one   C. another one   D. the others

6. Japan is _____ the east of China.

   A. in   B. to   C. on   D. at

7. – Do you speak English?

   – Yes, I speak _____ a little English _____ some French.

   A. Neither, not   B. both, and

   C. either, or   D. not only, but also

8. Janes _____ a new dress every month when she was in Shanghai.

   A. buys   B. is buying   C. bought   D. will buy

9. The population of the world in 20th century became very much _____ than that in 19th.

   A. bigger   B. larger   C. greater   D. more

10. This cake looks _____ to me, and I like it.

   A. terrible   B. good   C. badly   D. nicely

Section Ⅳ　Writing

Telephone Message（电话留言）

　　电话留言是指在对方无法接听电话时，留下的一种通信方式。电话留言通常用于传达重要信息，如请求、约会或提醒，并可在以后的时间检索和回复。留言条简明扼要，但又不能丢失信息，以便保证信息传达的完整性。

Part 1　Sample

The following is a telephone message. Please read and try to understand it.

Telephone Message	
Date：March 12th	**Time**：10：00 a. m.
From：Jane Smith	
To：Mr. John Doe	
Mobile phone number：555-1234	
Message：Jane Smith wanted to touch base regarding the upcoming meeting on Friday. He had a few questions and concerns that he wanted to discuss with you. He would be available until 4 p. m. today, and after 10 a. m. tomorrow.	
Signed by：Helen	

Part 2　Template

　　从上面的样例可以看出，英语电话留言条（Telephone Message）通常包括以下几个基本要素：

1. 发话者（From）
2. 接收电话留言者（To）
3. 日期（Date）
4. 时间（Time）
5. 发话者电话号码（Telephone number）
6. 留言内容（Message）
7. 有时还包括记录人签名（Signed by）

Part 3　**Useful Patterns**

1. Sb. has just rung up saying that…	1. 某人刚刚来电话说…
2. Sth. about…	2. 关于…
3. Please call sb. at (telephone number) about sth.	3. 请打电话给某人，关于某事，他/她的电话号码是…
4. Here is a message from sb. for you.	4. 这是某人给你的电话留言。
5. He said he would ring later again.	5. 他说他会再打电话过来。
6. Please ring him as soon as possible.	6. 请尽快给他回电。
7. I'm sorry, but he's on another line now.	7. 对不起，他在接另一个电话。
8. Anything else?	8. 还有其他事吗?
9. Just a minute. I'll get a pen.	9. 请等一下，我拿支笔。
10. I'll give him the message.	10. 我会转告他。

Part 4　**Exercises**

I. Translate the following telephone message into Chinese.

> **Telephone Message**
>
> Date: Dec. 1st
>
> Time: 11:00 a. m.
>
> From: David Brown
>
> Mobile phone number: 555-1234
>
> To: Dr. Smith
>
> Message: He wanted to follow up regarding the appointment you scheduled for the next week. Unfortunately, he needed to reschedule due to a conflict in his schedule. Could you please call him back at your earliest convenience to discuss alternative dates?
>
> Signed by: Li Ming

II. Write a telephone message according to the tape script of a telephone conversation.

电话内容:

A: Good morning. ABC Import and Export Company. Can I help you?

B: Yes, may I speak to Mr. Johnson, please?

A: I'm afraid Mr. Johnson isn't available right now. Would you like to leave a message?

B: This is Mary Green. It's very important that he should return my call this morning.

A: Does he have your office number and your mobile phone number?

B: I think so, but let me give them to you again.

A: Okay.

B: My office number is 000-111-222. My mobile phone number is 1311…1234. He can reach me at my office before 11: 00 a. m. or any time this afternoon on my mobile.

A: Very well, I'll give him your message as soon as he returns to the office.

Telephone Message

From: _____

Mobile phone number: _____

Office number: _____

To: _____

Message: _____

Unit 2

Computer Software

Unit Goals

After learning this unit, you will be able to:

- talk about computer software
- understand the passages and grasp the key words and expressions
- review the grammar of pronoun
- write a notice

Section Ⅰ Listening & Speaking

Linux

Mac OS X

Windows

Part A Words and Expressions

project 项目；工程；方案	migrate 转移
category 类别；(人或事物的) 种类	function 作用；功能
input 输入；投入	professional 专业的
automatically 自动地；机械地	schedule 工作计划；日程安排
insurance 保险	tight 紧的
data 数据；资料	assign 分派；布置 (工作、任务等)
interface 接口；接口程序	diagram 简图；图解
design 设计	draft 草稿；草案
install 安装	

Part B Conversations

Read the conversations carefully and then complete the communicative tasks.

Conversation 1

A：Glad to meet you, Susan. Welcome to XY Life! I'm Richard.

B：Great to see you, Richard!

A：I will be your responder and main resource throughout the whole **project**. If you need help at any time, just feel free to contact me.

B：I will. This is the requirements **category**. Briefly we need to know what the system should do and what the hardware/software constraints, etc.

A：Basically, the system is called FAS-Financial Analysis System. It's for our customers. After **inputting** their personal information such as age, gender, occupation and so on, the system will **automatically** show all our life **insurance** products that suit for them.

B：When can I get the products **data**?

A：In about 2 weeks. We've set up a team to categorize all the products data to make it easier to check.

B: Great! I feel much easier to discuss requirements with IT professionals like you.

A: Thank you. I've arranged a meeting at 9:00 tomorrow morning, so we can have a full discussion then.

B: That would be great.

A: See you tomorrow then.

 Key Words

project　项目；工程；方案

category　类别；（人或事物的）种类

input　输入；投入

automatically　自动地；机械地

insurance　保险

data　数据；资料

Task 1　Act out the conversation with your partner based on the following clues.

A: Glad to meet you, Susan. Welcome to XY Life! I'm Richard.

B: Great to see you, Richard!

A: I will be your responder and main resource throughout the whole project. ____①____.

B: I will. This is the requirements category. ____②____, etc.

A: Basically, the system is called FAS-Financial Analysis System. It's for our customers. After inputting their personal information such as age, gender, occupation and so on, the system will automatically show all our life insurance products that suit for them.

B: ____③____.

A: In about 2 weeks. We've set up a team to categorize all the products data to make it easier to check.

B: Great! ____④____.

A: Thank you. I've arranged a meeting at 9:00 tomorrow morning, so we can have a full discussion then.

B: ____⑤____.

A: See you tomorrow then.

Conversation 2

A: Morning, Susan.

B: Morning, Richard. Where should we start?

A: Let's begin with the **interface design**. But before that, I will show you something. Take this U-disk, **install** the software and run it. It's a FAS from another life insurance company, and we could use it as a model. We could **migrate** some **functions** and create some new for our own.

B: Great! That really makes the task much easier. You are really very **professional**.

A: Thank you.

 Key Words

interface 接口；接口程序
design 设计
install 安装
migrate 转移
function 作用；功能
professional 专业的

Task 2 Act out the conversation with your partner based on the following clues.

A: Morning, Susan.

B: Morning, Richard. _____①_____?

A: _____②_____.

But before that, I will show you something. Take this U-disk, _____③_____. It's a FAS from another life insurance company, and _____④_____. We could migrate some functions and create some new for our own.

B: Great! That really makes the task much easier. _____⑤_____.

A: Thank you.

Conversation 3

A: Everybody here? OK. First, good news for you, the contract with XY Life was signed yesterday. Kevin and I have started to **schedule** the project plan for the whole project and for each of you. Kevin, please.

B: Hi, guys, glad to work with you again! Time is **tight**. Later, I will present you the time table. Any questions?

C: Could you **assign** the task to each team?

B: OK! Henry team is responsible for the context model; Daniel team, the data-flow **diagrams**; my team, the object model. Susan, you work with the design team to provide 3 **draft** interface designs.

D: I got an e-mail from Richard this morning saying that there might be some changes, no more details, shall we wait?

A: No, we don't have time to wait, just go as scheduled.

Key Words

schedule　工作计划；日程安排

tight　紧的

assign　分派，布置（工作、任务等）

diagram　简图；图解

draft　草稿；草案

Task 3　Act out the conversation with your partner based on the following clues.

A： Everybody here? OK. First, good news for you, _____①_____ . Kevin and I have started to schedule the project plan for the whole project and for each of you. Kevin, please.

B： Hi, guys, _____②_____ ! _____③_____ . Later, I will present you the time table. Any questions?

C： Could you assign the task to each team?

B： OK! _____④_____ ; Daniel team, the data-flow diagrams; my team, the object model. Susan, you work with the design team to provide 3 draft interface designs.

D： I got an e-mail froZm Richard this morning saying that there might be some changes, no more details, shall we wait?

A： No, we don't have time to wait, _____⑤_____ .

| Part C | Passages |

Listen to the following passages carefully and fill in the blanks with the information you've heard.

听力材料

Passage 1

The fundamental _____ between hardware and software is that the former is tangible while the latter is not. The hardware is the _____ itself, doing all the physical work, while the software tells the various hardware components what to do and how to _____ . This enables the computer to adapt to new tasks or install new hardware. Hardware _____ display, central processing unit (CPU), keyboard and mouse; software includes operating system, word processing _____ and games, etc.

Passage 2

Application software is the part of software provided to meet the _____ needs of users in different _____ and different _____ . Running on the _____ system, it can broaden the application field of computer system and _____ the function of hardware.

Passage 3

The _____ computer did not have an operating system. People used _____ buttons to control the computer. Later, assembly language appeared. The operator entered the program into the computer through a paper tape with holes for compilation. These computers with built-in _____ can only be run by the producers themselves, which is not conducive to the sharing of programs and equipment. In order to _____ this problem, an operating system has emerged, which enables the sharing of programs and the _____ of computer hardware resources.

Passage 4

Application software is used to _____ a specific task, such as clearing accounts, processing _____ or drafting _____ . The classification of application programs includes: word processing programs, spreadsheet programs, presentation programs, database management software, graphics software, multimedia programs, e-mail programs, web browsers, _____ software, suite programs, accounting _____ , custom programs, etc.

Passage 5

This _____ of program allows users to extend the functionality of the graphic design program and add _____ controls similar to video, audio or animation to their own _____ . These programs are gradually becoming easy to use. _____ , multimedia controls are also added to document files published on the _____ or intranet.

Section II　Reading

Passage 1

Computer Software

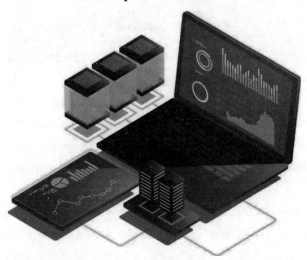

Computer **software** is essentially a program, which is a collection of computer data and instructions organized in a specific order. Since the program has no fixed material form, it can only function by installing it on a computer, so it is called software.

The **fundamental** difference between hardware and software is that the former is tangible while the latter is not. [1] The hardware is the machine itself, doing all the physical work, while the software tells the various hardware components what to do and how to interact. [2] This enables the computer to adapt to new tasks or install new hardware. Hardware includes display, central processing unit (CPU), **keyboard** and mouse; software includes operating system, word processing program and games, etc.

There are two main types of computer software: system software (operating system) and application software.

1. The most important and basic thing in computer software is the operating system (OS). It is the lowest level of software. It controls the programs run by all computers and manages the hardware resources of the entire computer. It is a bridge between computer bare metal and application programs and users. Without it, users cannot use some kind of software or program. The operating system is the control and management center of the computer system. From a **resource** point of view, it has 4 functions such as processor, memory **management**, device management, and file management. [3] **Commonly** used systems include WINDOWS operating system, MacOS operating system, UNIX operating system and Linux operating system.

2. Application software is the part of software provided to meet the application needs of users in different fields and different problems. Running on the operating system, it can **broaden** the application field of computer system and enlarge the function of hardware. ④

For example: office software, database software, programming software, Internet software, instant messaging software, e-mail client, web browser, multimedia software, etc.

3. Software license.

Different software generally has a **corresponding** software **license**, and the user of the software must agree to the license of the software used to use the software **legally**. On the other hand, the license terms of a particular software cannot conflict with the law. ⑤ Copies of software without the permission of the software copyright owner will cause legal problems. Generally speaking, it is also illegal to purchase and use this pirated software.

There is also some software that are open-source software developed by programmers and computer enthusiasts and can be **obtained** and used for free.

New Words and Expressions

software [ˈsɒftweə(r)] n. 软件

fundamental [ˌfʌndəˈmentl] adj. 根本的；基础的；基本的

keyboard [ˈkiːbɔːd] n. (计算机或打字机的)键盘

resource [rɪˈsɔːsɪz] n. 资源

management [ˈmænɪdʒmənt] n. 经营；管理

commonly [ˈkɒmənli] adv. 通常地；通常

broaden [ˈbrɔːdn] v. 变宽；变阔；增长

corresponding [ˌkɒrəˈspɒndɪŋ] adj. 符合的；相应的；相关的

license [ˈlaɪsns] n. 许可证；执照 vt. 许可

legally [ˈliːgəli] adv. 合法地；在法律上

enthusiast [ɪnˈθjuːzɪæst] n. 爱好者；热衷于…的人；热心者

obtain [əbˈteɪn] v. 获得；赢得

Notes

1. "The fundamental difference between hardware and software is that the former is tangible while the latter is not."

   Analysis："…the former is tangible while the latter is not." 表语从句，the former…，the latter…表示"前者…后者…"。

   Translation：硬件和软件之间的根本区别在于，前者是有形的，而后者则不是。

2. "The hardware is the machine itself, doing all the physical work, while the software tells the various hardware components what to do and how to interact."

   Analysis："while" 连词，表示转折，意为"然而"。"tell" 为及物动词，后接双宾语，一是 the various hardware components；二是 what to do and how to interact。

   Translation：硬件是机器本身，完成所有的物理工作，而软件告诉各种硬件组件要做什么以及如何相互作用。

3. "From a resource point of view, it has 4 functions such as processor, memory management, device management, and file management."

   Translation：从资源角度来看，它具有处理机、存储器管理、设备管理、文件管理4项功能。

4. "Running on the operating system, it can broaden the application field of computer system and enlarge the function of hardware."

   Analysis："Running on the operating system" 现在分词作状语，表示方式。

   Translation：运行在操作系统之上，它可以拓宽计算机系统的应用领域，扩大硬件的功能。

5. "On the other hand, the license terms of a particular software cannot conflict with the law."

   Analysis："on the other hand" 表示"在另一方面"通常与 on one hand 连用。

   Translation：从另一方面来讲，某种特定软件的许可条款也不能够与法律相抵触。

Exercises

I. Answer the following questions according to the text.

1. What is computer software?

   _____

   _____

2. What is the fundamental difference between hardware and software?

   _____

   _____

3. How many types of computer software? What are they?

_____

_____

4. Can you set some examples of software?

_____

_____

5. Is it legal or illegal to purchase and use a pirated software?

_____

_____

II. Fill in the blanks with words according to the meaning of the article by memory.

The most important and _____ thing in computer software is the operating system (OS). It is the lowest level of software. It _____ the programs run by all computers and manages the hardware resources of the entire computer. It is a _____ between computer bare metal and _____ programs and users. Without it, users cannot use some kind of software or program. The operating system is the control and management center of the computer _____. From a resource point of view, it has 4 functions such as processor, _____ management, device management, and file management. Commonly used systems _____ WINDOWS operating system, MacOS operating system, UNIX operating system and Linux operating system.

III. Fill in the blanks with the words given below. Change the forms when necessary.

since	meet	former
control	essentially	with
operate	legal	

1. The most important and basic thing in computer software is the _____ system.

2. It is also _____ to purchase and use this pirated software.

3. Application software is the part of software provided to _____ the application needs of users in different fields and different problems.

4. Copies of software _____ the permission of the software copyright owner will cause legal problems.

5. Computer software is _____ a program, which is a collection of computer data.

6. The fundamental difference between hardware and software is that the _____ is tangible while the latter is not.

7. _____ the program has no fixed material form, it can only function by installing it on a computer, so it is called software.

8. It _____ the programs run by all computers and manages the hardware resources of the entire computer.

IV. Translate the following sentences into Chinese.

1. Computer software is essentially a program, which is a collection of computer data and instructions organized in a specific order.

_____

_____

2. The most important and basic thing in computer software is the operating system.

_____

_____

3. Application software is the part of software provided to meet the application needs of users in different fields and different problems.

_____

_____

4. Different software generally has a corresponding software license, and the user of the software must agree to the license of the software used to use the software legally.

_____

_____

5. However, ads posted by a small branch of the company might only aim at getting students to fill up temporary vacancies in times of labor shortages.

_____

_____

Passage 2

Operating System

The original computer did not have an operating system. People used **various** buttons to control the computer. Later, assembly language appeared. The operator entered the program into the computer through a paper tape with holes for compilation. These computers with built-in languages can only run by the producers themselves, which is not conducive to the sharing of programs and equipment. In order to **solve** this problem, an operating system has emerged, which enables the sharing of programs and the management of computer hardware resources. [①]

An operating system or system environment is a collection of programs used to control human-computer interaction and **communication**. [②] It performs two important functions of the computer:

Manage input devices, output devices, storage devices, etc.

Manage file storage and **identification** to complete tasks.

Every computer needs an operating system to run. Computers without operating systems are called "**bare machine**". Computers can only install other applications on the basis of operating systems.

1. Desktop operating systems include：Windows, UNIX, Linux, MacOS, etc.

Windows or Mac OS are both graphical operating systems. Because this type of operating system makes it easy to apply computers, it becomes the standard of operating systems. With the advancement of technology, the display function has been strengthened, making the principle of using the screen to complete the design work simple. ③

The graphical user interface (GUI) allows users to use a mouse or other device to point to and select a command without having to remember the specific content of the command. These commands appear as buttons or pictures or symbols describing the command. The software provider also uses the same buttons/symbols/pictures in the designed program to complete the same functions (such as copy, paste, bold, save, print, etc.) in order to reduce the time spent by users learning to use new software.

The Windows of Microsoft is the operating system used by personal computers. Windows products support "what you see is what you get" screen displays and can preview documents **immediately** before printing.

MacOS is an operating system designed by Apple for iMAC computers. It provides users with a graphical interface, which makes it easy and fast for application computers to work, and truly sets standards for "what you see is what you get" programs. ④

The UNIX operating system was jointly developed by many programmers in the 1970s. The system is widely used in high-end servers. The main disadvantage of this operating system is that it implements control functions based on the command line, which is unfriendly to beginners.

The Linux operating system is developed based on UNIX technology and provides more graphical interfaces than UNIX. The system is widely used in mid-range servers or personal workstations, and is used by enterprise software developers.

2. The **mobile terminal** operating system includes Android, IOS, etc.

 New Words and Expressions

various [ˈveəriəs]　*adj.* 各种各样的；不同的

solve [sɒlv]　*v.* 解决；处理

communication [kəˌmjuːnɪˈkeɪʃn]　*n.* 表达；交流

identification [aɪˌdentɪfɪˈkeɪʃn]　*n.* 确定；确认；识别

baremachine　裸机

immediately [ɪˈmiːdiətli]　*adv.* 立即；马上

mobile [ˈməʊbaɪl]　*adj.* 可移动的

terminal [ˈtɜːmɪnl]　*n.* (电路的)端子；线接头

Notes

1. "In order to solve this problem, an operating system has emerged, which enables the sharing of programs and the management of computer hardware resources."

   Analysis："enable"动词，意为"使能够做…"。

   Translation：为了解决这种问题，操作系统出现了，这样就很好地实现了程序的共享和对计算机硬件资源的管理。

2. "An operating system or system environment is a collection of programs used to control human-computer interaction and communication."

   Analysis："used to control human-computer interaction and communication."过去分词做定语修饰"a collection of programs"。

   Translation：操作系统或系统环境是一个用来控制人机交互和通信的程序集合。

3. "With the advancement of technology, the display function has been strengthened, making the principle of using the screen to complete the design work simple."

   Translation：随着技术的进步，显示功能得到加强，使利用屏幕完成设计工作的原理变得简单。

4. "It provides users with a graphical interface, which makes it easy and fast for application computers to work, and truly sets standards for "what you see is what you get" programs."

   Analysis："provide sb. with sth."意为"给某人提供某物"。

   Translation：它为用户提供了一个图形化界面，这使应用计算机进行工作既容易又快速，并真正为"所见即所得"程序制定了标准。

Exercises

Answer the following questions according to the text.

1. What important functions of the computer does the operating system perform?

   _____

   _____

2. What do desktop operating systems include?

   _____

   _____

3. What is MacOS operating system?

   _____

   _____

4. What is the Linux operating system?

5. What does the mobile terminal operating system include?

Passage 3

Application Software

Application software is used to complete a specific task, such as clearing accounts, processing words or drafting documents. The classification of application programs includes: word processing programs, **spreadsheet** programs, **presentation** programs, **database** management software, graphics software, multimedia programs, e-mail programs, web browsers, tool software, suite programs, **accounting** programs, custom programs, etc.

1. Word Processing Software

Word processing software is the most common application software, which allows users to create, edit and save documents. When using a typewriter, the user may have to reprint the original document,[①] while when using a computer, the document can be stored electronically, and the user can easily find it again and modify it. Many professional word processing programs have typesetting functions.

2. Database Management Software

Simply put, a database is a collection of relevant information. Common examples are phone books, inventory lists, and personal files. Database Management Software (DMS) is used to help operate and manage information in the database.

In the past, people's demand for data storage was usually realized by filing cabinets, folders and other storage tools. No matter what kind of storage tools, they needed to be well organized and convenient for users to store information according to their needs. [②]

Users can generate reports or views on any field in a table through query commands, set up key fields, or establish connections between tables to generate various reports that can share information in tables in multiple databases. The working method of relational databases is that data between different databases can be associated with fields with the same name and share information about fields with the same name.

A database application software is to operate the information that needs to be managed and has the right to use in various necessary ways through archive cabinets, various types of folders or other information storage systems.

3. Multimedia Programs

This type of program allows users to extend the functionality of the graphic design program and add media controls similar to video, audio or animation to their own files. These programs are gradually becoming easy to use. Similarly, **multimedia** controls are also added to document files published on the Internet or intranet.

4. Web Browser

A web **browser** is a program that allows users to connect to the Internet and browse the websites of different companies, organizations or individuals. A large number of users begin to surf the Internet, as well as companies and individuals set up their own websites, browsing the Internet has become more and more popular. [③] Users can use addresses or domain names to switch between websites.

Microsoft's Microsoft Edge browser is a typical web browser. It is bundled with Windows 10 operating systems. The latest version can be obtained from Microsoft's website.

5. Tool Software

Anti-virus protection software, it is necessary to buy an anti-virus software for computers. Having an anti-virus program (and keeping it updated at any time) can protect the system from unnecessary computer viruses. For example, Jinshan's Jinshan Poison Bulls and 360's 360 killing kits are very popular anti-virus programs.

 New Words and Expressions

application [ˌæplɪˈkeɪʃn]   *n.* 应用

spreadsheet [ˈspredʃiːt]   *n.* 电子表格

presentation [ˌpreznˈteɪʃn]   *n.* 演示

database [ˈdeɪtəbeɪs]   *n.* 数据库

account [əˈkaʊnt]   *n.* 账户

inventory [ˈɪnvəntri]   *n.* 库存

multimedia [ˌmʌltiˈmiːdiə]   *n.* 多媒体   *adj.* 多媒体的

browser [ˈbraʊzə(r)]   *n.* 浏览器

anti-virus [ˈænti ˈvaɪrəs]   *adj.* (软件)杀毒的；防病毒的

 Notes

1. "When using a typewriter, the user may have to reprint the original document."
   Analysis："When using a typewriter…" 现在分词表示时间状语，意为 "当…时"。
   Translation：使用打字机时，用户可能不得不对原有文档进行重新打印。

2. "No matter what kind of storage tools, they needed to be well organized and convenient for users to store information according to their needs."
   Analysis："No matter what…" 引导让步状语从句，意为 "无论什么…"。
   Translation：无论哪类存储工具，都需要整理好并方便使用者按照需求存储信息。

3. "A large number of users begin to surf the Internet, as well as companies and individuals set up their own websites, browsing the Internet has become more and more popular."
   Analysis："a large number of" 意为 "许多；大量" 后接可数名词复数形式。
   Translation：大量用户开始上网，公司和个人纷纷建立自己的网站，浏览互联网已经变得越来越流行。

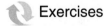 Exercises

Translate the following short passages into Chinese.

1. Application software is used to complete a specific task, such as clearing accounts, processing words or drafting documents. The classification of application programs includes：word

processing programs, spreadsheet programs, presentation programs, database management software, graphics software, multimedia programs, e-mail programs, web browsers, tool software, suite programs, accounting programs, custom programs, etc.

_____

_____

_____

2. This type of program allows users to extend the functionality of the graphic design program and add media controls similar to video, audio or animation to their own files. These programs are gradually becoming easy to use. Similarly, multimedia controls are also added to document files published on the Internet or intranet.

_____

_____

_____

3. Anti-virus protection software, it is necessary to buy an anti-virus software for computers. Having an anti-virus program (and keeping it updated at any time) can protect the system from unnecessary computer viruses. For example, Jinshan's Jinshan Poison Bulls and 360's 360 killing kits are very popular anti-virus programs.

_____

_____

_____

Section Ⅲ　Grammar

Pronoun （代词）

代词是代替名词、形容词和数词的词类。

一、代词的分类

英语中的代词，按其意义、特征及在句中的作用分为：人称代词、物主代词、指示代词、自身代词、相互代词、不定代词、疑问代词和关系代词八种。

（一）人称代词

人称代词是表示"我""你""他""她""它""我们""你们""他们"指代人或事物的代词。人称代词有人称、数和格的变化。见下表：

数	单数		复数	
格	主格	宾格	主格	宾格
第一人称	I	me	we	us
第二人称	you	you	you	you
第三人称	he	him	they	them
	she	her	they	them
	it	it	they	them

（二）物主代词

物主代词是表示所有关系的代词。物主代词分形容性物主代词和名词性物主代词两种，其人称和数的变化见下表：

数	单数			复数		
人称	第一人称	第二人称	第三人称	第一人称	第二人称	第三人称
形容词性物主代词	my	your	his/her/its	our	your	their
名词性物主代词	mine	yours	his/hers/its	ours	yours	theirs

（三）指示代词

指示代词是表示"这个""那个""这些""那些"指示概念的代词。可分单数（this/that）和复数（these/those）两种形式，既可作限定词又可做代词。见下表：

	单数	复数
限定词	This girl is Mary.	Those men are my teachers.
代词	This is Mary.	Those are my teachers.

（四）自身代词

自身代词是表示"我自己""你自己""他自己""我们自己""你们自己"和"他们自己"等的代词，也称为"反身代词"。见下表：

数	单数			复数		
人称	第一人称	第二人称	第三人称	第一人称	第二人称	第三人称
人称代词	I	you	he/she/it	we	you	they
反身代词	myself	yourself	himself/ herself/ itself	ourselves	yourselves	themselves

（五）相互代词

相互代词是表示相互关系的代词，有 each other 和 one another 两组，但在运用中，这两组词没什么大的区别。

例如：They love each other. 他们彼此相爱。

（六）不定代词

不定代词是表示不指明代替任何特定名词或形容词的代词。常见的不定代词有 all，both，each，every 等，以及含有 some-，any-，no-等的合成代词，如：anybody，something，no one。这些不定代词大都可以代替名词和形容词，在句中作主语、宾语、表语和定语，但 none 和由 some，any，no 等构成的复合不定代词只能作主语、宾语或表语；every 和 no 只能作定语。

（七）疑问代词：

疑问代词在句中起名词词组的作用，有 who，whom，whose，what 和 which 等，在句子中用来构成特殊疑问句。

例如：Who is that girl? 那个女孩是谁？

疑问代词还可以用作连接代词，引导名词性从句（主语从句、宾语从句和表语从句）。

例如：Tell me who he is. 告诉我他是谁。

（八）关系代词

关系代词有 who，whom，whose，that，which，as 等。一方面可用作引导定语从句的关联词，另一方面它们又代表主句中为定语从句所修饰的那个名词或代词（先行词），在定语从句中作主语、表语、宾语、定语等。

例如：He is the man whom you have been looking for. 他就是你要找的那个人。

二、不定代词的具体用法（见下表）

表一：

	某	任何	每个/所有	没有
人	somebody	anybody	everybody	nobody

续表

	某	任何	每个/所有	没有
人	someone	anyone	everyone	no one
物	something	anything	everything	nothing

1. some，any，no 和 every 可以和 one，body，thing 构成复合代词，它们的用法与 some，any 的用法相同。some 用在肯定句和期待得到对方肯定回答的疑问中；any 用于疑问句、否定句和条件中。

2. 修饰不定代词的词一定要放在不定代词后。

I have something important to tell you.

3. 当主句的主语是人的复合不定代词，如 everybody，nobody，anyone 等时，其反意疑问句的主语通常用代词 they；当主句中的主语是物的复合不定代词，如 everything，anything，something，nothing 等时，其反意疑问句的主语用代词 it.

Everything is ready, isn't it?

Everyone is here, aren't they?

4. 复合不定代词本身做主语时，谓语用单数。

5. everybody = everyone，只能指人；every one 既可指人又可指物，还可以和 of 连用。

表二:

	都	任何	都不
两者	both	either	neither
三者（以上）	all	any	none

1. both 两者都；all 三者或三者以上人或物都。在句中作主语时，谓语用复数。与 not 连用时，表示部分否定，译为"并非所有都…"

Both John and Ann have friends.

All of us don't like meat. = Not all of us like meat.

2. either 用于两者，指两者之一（两者间任何一个）；any 用于三者或三者以上，指任意一个。

When shall we meet, this evening or tomorrow morning? Either time is OK.

When shall we meet? Any time is OK.

3. neither 是 both 的反义词，表示"两者都不"。作主语时，谓语用单数。none 是 all 的反义词，表示"三者或三者以上的人或物都不。"作主语时，可单可复。注意 none 与 no one 的区别。

表三：

	指代	用来回答	相当于
none	人或物	how many/much	not a/an/any + 名词或 "no + 名词"
nobody/no one	人	who	not anyone/not anybody
nothing	物	what	not anything

none 后面可跟 of 短语，而 something/anything/everything/nothing 和 someone/anyone/everyone/no one 却不能。

None of them knew him. 没人认识他。

表四：

	可数	不可数
肯定	a few（有一些）	a little（有一点）
否定	few（几乎没有）	little（几乎没有）

There is little water in the bottle, is there?

There is a little water in the bottle, isn't there?

表五：

不定代词	意义	用法说明
other	另外的	只作定语，常与复数名词或不可数名词连用；前面有 the, this, that, some, your 等时，则可以与单数连用。Other + 名词 = others
the other	两者中的另一个	常与 one 连用，构成 one...the other；the other + 名词 = the others
others	泛指别的人或物	是 other 复数形式；泛指别的人或物（但不是剩余的全部），不能作定语，构成 some...others...
the others	特指其余的人或物	是 the other 的复数形式，特指剩余的全部的人或物
another	任何一个，另一个	指三者或三者以上中的任何一个，用作形容词或代词 another + 单数名词；another + 数词 + 复数名词 译为 "再，又"

例句：

1. I have two pens. **One** is red, the **other** is blue.

2. I don't like the book, please give me **another** one.

3. **Some** students are listening to the radio, **others**（= other students）are watching TV.（可能还有一些人在干别的。）

4. There are thirty students in our class. Twenty are girls, **the other** students（= the others）are boys.

5. There are thirty students in our class. Twenty are from Beijing. **The others** are from

Nanjing.

（其余的全来自南京）

6. There are thirty students in our class. Twenty are from Beijing. **Others** are from Nanjing. （还有来自南京的，暗示其余十人不只是来自南京，还有其他地方来的。）

7. You can stay **another ten days**. = You can stay **ten more days**.

Exercises

Choose the best answer.

1. – Mary, is this _____ blue bike?

　 – No, _____ is a black one.

　 A. your, my　　　B. your, mine　　　C. yours, my　　　D. yours, mine

2. – Do we have any bananas for the party?

　 – Let me see. Yes, but only _____ .

　 A. a few　　　　B. few　　　　C. a little　　　　D. little

3. "Help _____ to some chicken," my mother said to the guests.

　 A. yourself　　　B. yourselves　　　C. your　　　D. yours

4. – If you are buying today's *Suzhou Daily*, could you get _____ for me?

　 – I'm glad to help you.

　 A. one　　　　B. it　　　　C. /　　　　D. this

5. _____ passed the exam.

　 A. You, I and she　　　　　　　　B. You, she and I

　 C. I, You and she　　　　　　　　D. She, I and you

6. – Is _____ here?

　 – No, Bob and Tim have asked for leave.

　 A. anybody　　　B. somebody　　　C. everybody　　　D. nobody

7. – When would you like to go to the park with me, this Friday or Saturday?

　 – _____ . I am free only this Sunday.

　 A. Both　　　　B. None　　　　C. Neither　　　　D. Either

8. Though I agree with most of what you said, that doesn't mean I agree with _____ .

　 A. something　　　B. anything　　　C. nothing　　　D. everything

9. The students have _____ on Sunday.

　 A. not any class　　B. not class　　C. no classes　　D. no any class

10. Mike is stronger than _____ in his class.

　 A. any boys　　　B. any boy　　　C. any other boys　　　D. other boy

Section IV Writing

Notice（通知）

通知是书信的一种形式，是上级对下级、组织对成员或平行单位之间部署工作、传达事情或召开会议等所使用的应用文。

通知有两种形式，一种是书信方式，寄出或发出，通知有关人员，此种通知写作形式同普通书信；另一种是布告形式，张贴通知。这里所讲的就是布告形式的通知。

Part 1 Sample

The following is a notice. Please read and try to understand it.

NOTICE

We're going to have interesting activities in the school library at 8:00 a. m. on November 20, 2002. Some of us will read poems and some will tell stories. You can also hear wonderful songs and watch beautiful dances there. We hope all students can come and join in the activities. All the headmasters will be invited to our activities as representatives of teachers. Please get one performance ready because some of you will probably be asked to give us one.

Students of Class 3, Senior 2

November 11, 2002

Part 2 Template

从上面的样例可以看出英文通知（Notice）的写作要领（Elements of Writing）如下：

1. 标题：通知上方居中写上标题，即 Notice，常大写为 NOTICE。

2. 日期：写在正文的右下方，在单位名称的下一行。

3. 称呼：一般无称呼语，而用第三人称来写。

4. 正文：会议或活动通知应写明时间、地点、活动内容、出席对象及有关注意事项。

5. 落款：出通知的单位或责任人名字写在正文最后的右下方，或放在标题之上，作为标题的部分。

6. 写通知时注意格式，内容完整，简单明了，中心突出。

NOTICE

Body of the notice _____

_____

_____

Signature:

Date:

Part 3　Useful Patterns

1. We'll meet at the school gate at 8:00 a. m.	1. 我们 8 点钟在校门口集合。
2. Don't be late.	2. 不要迟到。
3. I have a piece of good news to tell you.	3. 我有个好消息要告诉大家。
4. Please take part in all the activities on time.	4. 请准时参加活动。
5. Please bring your lunch by yourselves.	5. 请大家自备午饭。
6. I have an announcement to make.	6. 我有事情要宣布。
7. Don't forget to come on time.	7. 别忘了准时出席。
8. Three things for attention.	8. 请大家注意三件事情。
9. This is to announce that…	9. 兹通知…

Part 4　Exercises

Ⅰ. **Translate the following notice into Chinese.**

NOTICE

The sports meeting whichis to take place this Saturday has to be put off because of the heavy rain these days. All students are required to come to school on Saturday morning as usual, but there will be no class that afternoon. Weather permitting, the sports meeting will be held next Saturday morning.

Office of Physical Education

10th September, 2009

Ⅱ. **Write a notice according to the information below.**

通　　知

本周星期五下午 4 点在团委办公室召开全体学生干部会议。会议重要，请勿迟到缺席。

学生会

2005. 3. 28

Unit 3
Computer Programming Language

After learning this unit, you will be able to:

- talk about computer programming languages
- understand the passages and grasp the key words and expressions
- review the grammar of simple sentence
- write a letter of thanks

Section I   Listening & Speaking

Part A   Words and Expressions

programming 编程	certificate *n.* 证明，证书；文凭
program 程序	PC = personal computer 个人计算机
Python 一种程序语言	process *v.* 处理
code 代码	computer processor 计算机处理机
fix 修复	qualification *n.* 资格
bug 故障；程序错误	rank *n.* 级别
data entry operator 数据录入员	assistance *n.* 帮助；援助；支持
NCRE 全国计算机等级考试	software *n.* （计算机）软件

Part B   Conversations

Read the conversations carefully and then complete the communicative tasks.

Conversation 1

A：Can I use this? I need to check my e-mail.

B：Sorry. I still need it for a while.

A：But you're just playing a game.

B：No, I'm not. I'm doing work for my code camp. Right now, I'm learning to make a game in Python.

A：What's Python?

B：It's a programming language. It's the easiest language, so it's the first one in my code camp.

A：Interesting. Can I test your game when it's finished?

B：Sure! That would help me a lot.

A：So what should I do?

B：You need to find a coin. To look for the coin, type "look". To go to the next room, type "leave". In one room, there's a dragon, so be careful!

A：Hmm. I typed "look", but nothing happened.

B：Oh, you found a bug in my code. I need to fix that later.

A：Do you know how?

B：Finding bugs is hard. Fixing them is easier.

A：Can you show me?

B：Yeah. If you're interested, I'll even teach you some simple Python. Then you can write your own game.

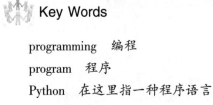

Key Words

programming　编程

program　程序

Python　在这里指一种程序语言

code　代码

code camp　代码训练营

fix　修复

bug　故障；程序错误

Task 1　Act out the conversation with your partner based on the following clues.

A：Can I use this? I need to check my e-mail.

B：Sorry. ____①____

A：But you're just playing a game.

B：No, I'm not. ____②____ Right now, I'm learning to make a game in Python.

A：What's Python?

B：____③____ It's the easiest language, so it's the first one in my code camp.

A：Interesting. Can I test your game when it's finished?

B：Sure! ____④____

A：So what showld I do?

B：You need to find a coin. To look for the coin, type "look". To go to the next room, type "leave". In one room, there's a dragon, so be careful!

A：Hmm. I typed "look", but nothing happened.

B：____⑤____ I need to fix that later.

A：Do you know how?

B：Finding bugs is hard. Fixing them is easier.

A：Can you show me?

B: Yeah. If you're interested, I'll even teach you some simple Python. Then you can write your own game.

Conversation 2

A: Do you have any experience working with a computer?

B: Yes, I have been a data entry operator for three years.

A: What kind of software can you use?

B: I have working knowledge of Windows and DOS. Actually, I'm quite familiar with both Java and C++ Programming Languages.

A: Do you have any other computer qualifications?

B: Yes. I have an NCRE certificate Rank 2.

A: Do you know how to use a PC to process the management information?

B: I'm sorry to say I'm not familiar with processing management information, but I'm sure I could learn quite quickly. It can't be too difficult, and I've got a quick mind. I can handle any problem you give me.

 Key Words

data entry operator   数据录入员

qualification   *n.* 资格

NCRE   全国计算机等级考试;

　　　　= National Computer Rank Examination

certificate   *n.* 证明,证书;文凭

rank   *n.* 级别

PC = personal computer   个人计算机

process   *v.* 处理

Task 2   Act out the conversation with your partner based on the following clues.

A: _____①_____

B: Yes, I have been a data entry operator for three years.

A: _____②_____

B: I have working knowledge of Windows and DOS. Actually, I'm quite familiar with both Java and C++ Programming Languages.

A: _____③_____

B: Yes. I have an NCRE certificate. Rank 2.

A: _____④_____

B: I'm sorry to say I'm not familiar with processing management information, but I'm sure I could learn quite quickly. _____⑤_____ I can handle any problem you give me.

Conversation 3

A： What is computer science?

B： The essence of computer science is to, with assistance of computer processor and software, solve real-world problems. Simply put, Computer Science is a school of study that studies the knowledge of computer programming and software.

A： What are the most popular programming languages?

B： For the time being, Java, C ++ , and Python are probably the three most popular programming languages globally.

A： What can you do with computer programming?

B： Just like the essence of computer science, you can solve many real-world problems by means of coding. For example, you can design an APP to help you choose the most suitable career in the future.

 Key Words

assistance　*n.* 帮助；援助；支持

computer processor　计算机处理机

software　*n.*（计算机）软件

Task 3　Act out the conversation with your partner based on the following clues.

A： _____①_____

B： The essence of computer science is to, with assistance of computer processor and software, solve real-world problems. Simply put, Computer Science is a school of study that studies the knowledge of computer programming and software.

A： _____②_____

B： For the time being, Java, C ++ , and Python are probably the three most popular programming languages globally.

A： _____③_____

B： Just like the essence of computer science, you can solve many real-world problems by means of coding. For example, you can design an APP to help you choose the most suitable career in the future.

Part C   Passages

听力材料

Listen to the following passages carefully and fill in the blanks with the information you've heard.

Passage 1

Computer _____ is a language used to _____ for the computer. The purpose of a programming language is to allow the _____ to express data processing activities in a _____ without regard to the details that the machine has to know. Programming languages fall into two major categories: _____ languages and high-level languages.

Passage 2

For the time being, _____, C++, and Python are probably the three most _____ programming languages globally. Each of them has some advantages and disadvantages. _____, Java and Python are easier to learn, with simpler formats and syntax. Meanwhile, C++ is more time-efficient and closer to the _____ system. Normally, it is highly _____ for starters to learn Java and Python first.

Passage 3

Java is a widely used _____ programming language and software _____ that runs on billions of _____, including _____ computers, mobile devices, gaming consoles, medical devices and many others. The rules and _____ of Java are based on the C and C++ languages.

Passage 4

It's also _____ to understand that Java is much _____ from JavaScript. JavaScript does not need to be _____, while _____ does need to be compiled. Also, JavaScript only runs on web _____ while Java can be run anywhere.

Passage 5

Learning programming is a way of _____ how you think. In order to do any good, you have to start thinking _____ and become more _____. If a program doesn't work, there is always a reason. It is up to you to either _____ out what's wrong and fix it or come up with a new method to _____ your goal.

Section II　Reading

Passage 1

Computer Programming Language

Computer **programming language** is a language used to write **instructions** for the computer. The purpose of a programming language is to allow the programmer to **express data** processing activities in a **symbolic** manner without regard to the details that the machine has to know. [1] Programming languages fall into two major categories: low-level **assembly languages** and high-level languages. [2] Assembly languages are **available** for every CPU family, and they translate one line of **code** into one machine language instruction. High-level languages translate programming statements into several machine instructions.

Normally, when people are talking about programming language, they are referring to those HLLs, or High Level Languages. **For instance**, Java, Python, C++ are all high-level languages. [3] Basically, high level languages are human-readable and human-understandable programming languages. To be more specific, in Java, if you want to print something like "I love computer science!" out on the **console**, you will just need to type "System. out. println（"I love computer science!"）". [4] Then your computer will totally understand this in Java and print it out on the screen for you to see. Machine language, on the other hand, is literally made of ones and zeroes, such as "10101011" and "00001". Your computer is able to directly understand the machine language. However, HLL will need to be compiled and converted to machine language in order for computers to work.

For the time being, Java, C++, and Python are probably the three most popular programming languages globally. ⑤ Each of them has some advantages and disadvantages. For instance, Java and Python are easier to learn, with simpler **formats** and **syntax**. Meanwhile, C++ is more time-efficient and closer to the **operating system**. Normally, it is highly advisable for starters to learn Java and Python first.

Just like the essence of computer science, you can solve many real-world problems by means of coding. For example, you can design an APP to help you choose the most suitable career in the future, or you can write some code to solve a hard puzzle, or you can simply write some code to find out the shortest **path** out from a certain **maze**. Anyway, programming makes your life easier.

New Words and Expressions

programming language　程序设计语言
instruction [ɪnˈstrʌkʃn]　n. 指示；教学；用法说明；计算机指令
express [ɪkˈspres]　v. 表达，表示
data [ˈdeɪtə]　n. 数据；资料
symbolic [sɪmˈbɒlɪk]　adj. 象征的，象征性的
assembly language　汇编语言
available [əˈveɪləbl]　adj. 可用的，可获得的；有空的
code [kəʊd]　n. 密码，代码；编码
for instance　例如；比如

statement ['steɪtmənt]　*n.* 说明；声明

console [kən'səul, 'kɒnsəul]　*v.* 安慰，慰问　*n.* (机器)控制台

format ['fɔ:mæt]　*n.* (出版物的)版式；[自] (数据安排的)形式

syntax ['sɪntæks]　*n.* 句法；句法规则[分析]；语构；语法

operating system　操作系统

path [pɑ:θ]　*n.* 小路；路线，道路；行动计划

maze [meɪz]　*n.* 迷宫；迷惑；错综复杂；迷宫图

Notes

1. The purpose of a programming language is to allow the programmer to express data processing activities in a symbolic manner without regard to the details that the machine has to know.

   Analysis："data processing activities" 意为"数据的处理活动"；"in a symbolic manner" 意思是"以符号的方式"。

   Translation：程序设计语言的目的是让程序设计者在不考虑机器是否能识别细节的情况下，以符号的方式表达对数据的处理活动。

2. Programming languages fall into two major categories：low-level assembly languages and high-level languages.

   Analysis："fall into" 这里意为"分成…"；"categories" 是 "category" 的名词复数，"种类" 的意思。

   Translation：程序设计语言主要分成两类：低级汇编语言和高级语言。

3. For instance, Java, Python, C++ are all high-level languages.

   Analysis："for instance" 的意思是"例如，比如"。

   Translation：比如 Java、Python、C++ 都是高级编程语言。

4. To be more specific, in Java, if you want to print something like "I love computer science!" out on the console, you will just need to type "System. out. println（"I love computer science!"）".

   Analysis："to be more specific" 意思是"具体而言，准确来说"。例如：I asked him to be more specific. 我要求他说得更具体些。

   Translation：更具体来说，拿 Java 来举个例子，如果你想在屏幕上打出"我热爱计算机科学!"这行字，你只需要输入 "System. out. println（"我热爱计算机科学!"）"。

5. For the time being, Java, C++, and Python are probably the three most popular programming languages globally.

   Analysis："for the time being" 意思是"目前"。例如：The situation is stable for the time being. 局势暂时稳定。

   Translation：目前来说，Java、C++、Python 很可能是最流行的三种计算机编程语言。

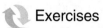 Exercises

I . Answer the following questions according to the text.

1. What is computer programming language?

_____

_____

2. What's the purpose of a programming language?

_____

_____

3. How many categories do programming languages have? What are they?

_____

_____

4. Can you tell us three most popular programming languages nowadays?

_____

_____

5. What are the advantages of C++?

_____

_____

II . Fill in the blanks with words according to the meaning of the article by memory.

Normally, when people are talking about _____ language, they are referring to those High Level Languages. For instance, Java, Python, C++ are all _____ languages. Basically, high level languages are human-readable and human-understandable programming languages. To be more specific, in Java, if you want to print something like "I love computer science!" out on the _____, you will just need to type "System. out. println ("I love computer science!")" . Then your _____ will totally understand this in Java and print it out on the _____ for you to see. Machine language, on the other hand, is literally made of _____ and _____, such as "10101011" and "00001" . Your computer is able to directly understand the _____ language. However, HLL will need to be compiled and converted to machine language in order for computers to work.

III. Fill in the blanks with the words given below. Change the forms when necessary.

refer	operate	direct
process	program	available
instruct	suit	

1. Computer _____ language is a language used to write instructions for the computer.

2. The purpose of a programming language is to allow the programmer to express data _____ activities in a symbolic manner.

3. Normally, when people are talking about programming language, they are _____ to those High Level Languages.

4. Your computer is able to _____ understand the machine language.

5. C++ is more time-efficient and closer to the _____ system.

6. Assembly languages are _____ for every CPU family.

7. High-level languages translate programming statements into several machine _____ .

8. You can design an APP to help you choose the most _____ career in the future.

IV. Translate the following sentences into Chinese.

1. Computer programming language is a language used to write instructions for the computer.

_____

_____

2. Programming languages fall into two major categories: low-level assembly languages and high-level languages.

_____

_____

3. Basically, high level languages are human-readable and human-understandable programming languages.

_____

_____

4. For the time being, Java, C++, and Python are probably the three most popular programming languages globally.

_____

_____

5. For instance, Java and Python are easier to learn, with simpler formats and syntax.

_____

_____

_____

_____

Passage 2

What is Java?

Java is a widely used **object-oriented** programming language and software **platform** that runs on billions of devices, including notebook computers, mobile devices, gaming consoles, medical devices and many others. ① The rules and syntax of Java are based on the C and C++ languages.

One major advantage of developing software with Java is its **portability**. Once you have written code for a Java program on a notebook computer, it is very easy to move the code to a mobile device. ② When the language was invented in 1991 by James Gosling of **Sun Microsystems** (later acquired by Oracle), the primary goal was to be able to "write once, run anywhere." ③

It's also important to understand that Java is much different from JavaScript. Javascript does not need to be **compiled**, while Java code does need to be compiled. Also, Javascript only runs on web **browsers** while Java can be run anywhere.

Java is a technology consisting of both a programming language and a software platform. To create an **application** using Java, you need to **download** the Java Development Kit (JDK), which is available for Windows, MacOS, and Linux. ④ You write the program in the Java programming language, then a compiler turns the program into Java **bytecode**—the instruction set for the Java **Virtual Machine** (JVM) that is a part of the Java runtime environment (JRE). ⑤ Java bytecode runs without **modification** on any system that supports JVMs, allowing your Java code to be run anywhere.

The Java software platform consists of the JVM, the Java API, and a complete development environment. The JVM parses and runs (interprets) the Java bytecode. The Java API consists of an extensive set of libraries including basic objects, networking and security functions; Extensible Markup Language (XML) generation; and web services. Taken together, the Java language and the Java software platform create a powerful, proven technology for enterprise software development.

New Words and Expressions

object-oriented [ˈəbdʒektˈɔːrɪəntɪd] *adj.* [计] 面向对象的

platform [ˈplætfɔːm] *n.* 计算机平台

portability [ˌpɔːtəˈbɪlətɪ] *n.* 可携带，轻便

Sun Microsystems 太阳微系统公司总部所在地

compile [kəmˈpaɪl] *v.* 收集；编辑；编译

browser [ˈbrəuzə] *n.* 浏览程序；浏览器(用于在互联网上查阅信息)

application [ˌæplɪˈkeɪʃn] *n.* 申请；申请书；应用软件

download [ˌdaʊnˈləʊd, ˈdaʊnləʊd] *v.* 下载

Virtual Machine 虚拟计算机；虚拟机

bytecode [bɪtiːˈkəʊd] *n.* 字节码

modification [ˌmɒdɪfɪˈkeɪʃn] *n.* 修改，修正

extensible [ɪkˈstensəbl] *adj.* 可展开的，可扩张的，可延长的

Notes

1. Java is a widely used object-oriented programming language and software platform that runs on billions of devices, including notebook computers, mobile devices, gaming consoles, medical devices and many others.

   Translation：Java 是一种广泛使用的面向对象编程语言和软件平台，可在数十亿设备上运行，包括笔记本电脑、移动设备、游戏控制台、医疗设备和许多其他设备。

2. Once you have written code for a Java program on a notebook computer, it is very easy to move the code to a mobile device.

Translation：一旦您在笔记本电脑上编写了 Java 程序的代码，就可以很容易地将代码移动到移动设备上。

3. When the language was invented in 1991 by James Gosling of Sun Microsystems（later acquired by Oracle）, the primary goal was to be able to "write once, run anywhere."

   Translation：1991 年，Sun Microsystems 的詹姆斯·戈斯林（James Gosling）（后来被甲骨文收购）发明了该语言，当时的主要目标是"只写一次，随时随地运行"。

4. To create an application using Java, you need to download the Java Development Kit（JDK）, which is available for Windows, macOS, and Linux.

   Translation：要使用 Java 创建应用程序，您需要下载 Java Development Kit（JDK），它适用于 Windows、MacOS 和 Linux。

5. You write the program in the Java programming language, then a compiler turns the program into Java byte code—the instruction set for the JavaVirtual Machine（JVM）that is a part of the Java runtime environment（JRE）.

   Translation：您用 Java 编程语言编写程序，然后编译器将程序转换为 Java 字节码——Java 虚拟机（JVM）的指令集，JVM 是 Java 运行时环境的一部分。

Exercises

Answer the following questions according to the text.

1. What is Java?

_____

_____

2. What is themajor advantage of developing software with Java?

_____

_____

3. What should you do first when you create an application using Java?

_____

_____

4. What does the Java software platform consist of?

_____

_____

5. Can you say something about the Java API?

_____

_____

Passage 3

What is Scratch?

Scratch is **agraphical** programming tool **released** by the "Lifelong Kindergarten Group" at MIT in 2007, and it is the most well-known one among all graphical programming tools. [1] You can use Scratch to write your own **interactive** stories, games, and **animations**, and share your creations with others in online **communities**.

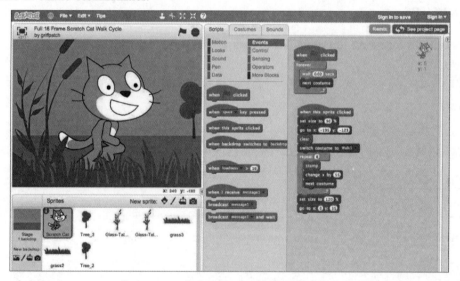

Scratch helps young people improve their creative thinking skills, logical reasoning skills, and project **collaboration** skills, which are all important skills for life in the 21st century. [2] Scratch is used by people in all countries of the world, and it is specifically designed for young people from 8 to 16 years old, but people of all ages can create and share Scratch programs (users range from children to the elderly).

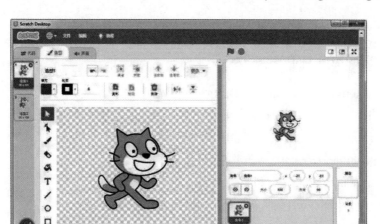

Instead of writing down line upon line of texts, you can drag "blocks" of code onto a board, which makes it nicely organized and easy to read, and complete the order. [3] Here are three steps to show how simple Scratch is:

1. Drag and drop;

2. Choose types of blocks;

3. Snap blocks together to create a script. [4]

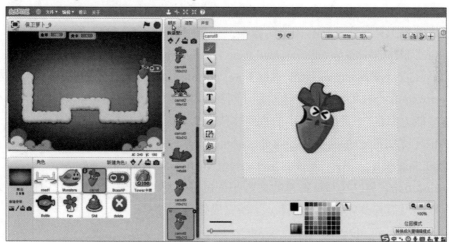

Another great thing about Scratch is that it is completely free. All you have to do is to look up "Scratch. mit. edu" on a **web browser** and create an account. It gets even better because it's also completely free to share your project with the world, so other people can try out your projects, give feedback on them, and even improve them![5]

Learning programming can change how you think. In order to do any good, you have to start thinking **creatively** and become more **adaptive**. If a program doesn't work there is always a reason. It is up to you to either figure out what's wrong and fix it or come up with a new method to achieve your goal. By identifying the problem in a program and coming up with a solution, you start to think more adaptively. By coming up with new methods to achieve a goal, you start thinking more creatively. This means that no matter what problems you come across, programming will always be able to improve your life skills.

 ## New Words and Expressions

graphical [ˈɡræfɪkl] *adj.* 绘成图画似的，绘画的

release [rɪˈliːs] *v.* 释放；发布；发行

interactive [ˌɪntərˈæktɪv] *adj.* 互相作用的，相互影响的；[计]交互式的；互动的

animation [ˌænɪˈmeɪʃn] *n.* 动画片（制作）

community [kəˈmjʊnəti] *n.* 团体，社区

collaboration [kəˌlæbəˈreɪʃn] *n.* 协作，合作

web browser 网页浏览器

creatively [krɪˈeɪtɪvlɪ] *adv.* 创造性地，有创造力地

adaptive [əˈdæptɪv] *adj.* 适应的；有适应能力的

Notes

1. Scratch is a graphical programming tool released bythe "Lifelong Kindergarten Group" at MIT in 2007, and it is the most well-known one among all graphical programming tools.
   Translation：Scratch 是麻省理工学院的"终身幼儿园团队"在 2007 年发布的一种图形化编程工具，它是所有图形化编程工具当中最广为人知的一种。

2. Scratch helps young people improve their creative thinking skills, logical reasoning skills, and project collaboration skills, which are all important skills for life in the 21st century.
   Translation：Scratch 帮助年轻人提升创新思维能力、逻辑推理能力和项目协作能力，这些能力都是 21 世纪生活中的重要技能。

3. Instead of writing down line upon line of texts, you candrag "blocks" of code onto a board, which makes it nicely organized and easy to read, and complete the order.
   Translation：与以往写一排排的文本不同的是，你可以把"代码块"拖到一块板上，这样代码很有条理并易于阅读，从而完成指令。

4. Snap blocks together to create a script.
   Translation：将块对齐以创建脚本。

5. It gets even better because it's also completely free to share your project with the world, so other people can try out your projects, give feedback on them, and even improve them!
   Translation：它还能更好，因为你还可以完全自由地与世界分享你的作品，这样其他人可以测试你的作品、对其进行反馈，甚至改进它们！

Exercises

Translate the following short passages into Chinese.

1. Scratch is used by people in all countries of the world, and it is specifically designed for young people from 8 to 16 years old, but people of all ages can create and share Scratch programs

(users range from children to the elderly).

_____

_____

_____

2. Another great thing about Scratch is that it is completely free. All you have to do is to look up "Scratch. mit. edu" on a web browser and create an account. It gets even better because it's also completely free to share your project with the world, so other people can try out your projects, give feedback on them, and even improve them!

_____

_____

_____

3. By identifying the problem in a program and coming up with a solution, you start to think more adaptively. By coming up with new methods to achieve a goal, you start thinking more creatively. This means that no matter what problems you come across, programming will always be able to improve your life skills.

_____

_____

_____

Section Ⅲ Grammar

Simple Sentence（简单句）

英语中，只含有一个主谓结构并且句子各成分都只由单词或短语构成的独立句子或分句叫作简单句。简单句的基本形式是由一个主语加一个谓语构成。其他各种句子形式都是由此发展而来，如五大基本句式结构：

1. S + V　主语 + 谓语

这种句式结构简称为主谓结构，其谓语一般都是不及物动词。例如：

Things change. 世事万变。

She cried bitterly. 她哭得很伤心。

2. S + V + P　主语 + 连系动词 + 表语

这种句式结构称为主系表结构，其实连系动词在形式上也是一种谓语动词，但实质上表语成了谓语。例如：

These books are novels. 这些书是小说。

The milk went sour. 牛奶变酸了。

She became a lawyer. 她当了律师。

Her voice sounded familiar. 她的声音听起来很熟悉。

The future seems hopeful. 未来似乎充满希望。

3. S + V + O　主语 + 谓语 + 宾语

这种句式结构可称为主谓宾结构，它的谓语一般多是及物动词。例如：

We never beat children. 我们从来不打孩子。

My sister will fix everything. 我姐姐会料理一切。

Angelia earned a score. 安洁莉亚取得了分数。

4. S + V + O1 + O2　主语 + 谓语 + 宾语1 + 宾语2

这种句式结构可称为主谓双宾结构，其谓语应是可有双宾语的及物动词。两个宾语一个是间接宾语，一个是直接宾语，其中指物或指事的就是直接宾语，指人（或动物）的就是间接宾语。例如：

He gave the book to his sister. 他把这本书给了他的妹妹。（book 是直接宾语，sister 是间接宾语）

I'll write you a long letter. 我将给你写一封长信。（letter 是直接宾语，you 是间接宾语）

He showed me his pencil-case. 他给我看他的铅笔盒。（pencil-case 是直接宾语，me 是间接宾语）

Tom gave his mom a present. Tom 送给妈妈一个礼物。（present 是直接宾语，mom 是间接宾语）

5. S + V + O + C　主语 + 谓语 + 宾语 + 宾补

所谓宾语补足语就是补充说明前面宾语的。这种句式结构可简称为主谓宾补结构，其补语是宾语补语，与宾语一起即构成复合宾语。例如：

I found the book easy. 我发现这本书不难。（形容词 easy 作补语）

I'll let him go. 我会让他去。（不定式 go 作补语）

John wants his brother to stay. 约翰让弟弟留下。（to stay 作补语）

The work made me exhausted. 这项工作使我筋疲力尽。（形容词 exhausted 作补语）

注意：有时两个或更多的并列主语拥有一个共同的谓语，甚至并列有两个主语和两个谓语，这样的句子仍然是简单句。例如：

China and other countries in the east Asia are developing rapidly. 中国和东亚其他国家正在迅速地发展。（China and other countries 为并列主语）

Mr. Wang and I often work together and help each other. 王先生和我常在一起工作互相帮助。

 Exercises

Choose the best answer.

1. As a good girl, you can't make your mother _____.

　　A. angrily　　　　B. angry　　　　C. happy　　　　D. happily

2. The apple tasted _____.

　　A. sweets　　　　B. sweetly　　　　C. nicely　　　　D. sweet

3. The actor _____ at the age of 70.

　　A. dead　　　　B. died　　　　C. dyed　　　　D. death

4. _____ were all very tired, but none of _____ would stop to take a rest.

　　A. We, us　　　　B. Us, we　　　　C. We, our　　　　D. We, we

5. The dog _____ mad.

　　A. looks　　　　B. looked　　　　C. is being looked　　　D. was looked

6. They will _____ Guangzhou tonight.

　　A. arrive　　　　B. get　　　　C. reach　　　　D. go

7. The recycled water（再生水）_____ harm to us.

　　A. do　　　　B. is　　　　C. doesn't do　　　　D. isn't

8. My teammates all call me _____.

　　A. captain　　　　B. to captain　　　　C. the captain　　　　D. to the captain

9. Tomorrow is Terry's birthday, so I will buy a doll _____ her.

　　A. for　　　　B. to　　　　C. on　　　　D. with

10. Jack is a good teacher; he _____.

　　A. teaches us English　　　　　　B. teaches our English

　　C. teaches us learn English　　　　D. teaches our to learn English

Section Ⅳ Writing

Letter of Thanks（感谢信）

感谢信是对收信人的某一行为表示感谢。在日常生活和工作中，感谢信应用范围很广。当收到馈赠、邀请、赴宴、照顾、款待、慰问、吊唁、祝贺时，写信道谢乃人之常情。在商务活动中，对雇员的建议，客户的订货，供应商的优质服务，银行、商会或友邻的信息提供等，均应写信致谢。感谢信能帮助加强联系、增进友谊、树立企业的良好形象。

Part 1 Sample

The following is a letter of thanks. Please read and try to understand it.

Dear Prof. Smith,

Many thanks for your hospitality and the honor you showed me during my recent visit to your country.

It was thoughtful of you to introduce me to so many of your famous professors and scholars in your country. I have learned a lot from them. I returned to China safe and sound last Friday, and have resumed my work now. I hope some day you will visit our country with its ancient history and beautiful scenery, and give us some lectures on "Modern English Literature".

Thank you for your many kindness to me. I hope to meet you soon.

Yours sincerely,

Li Yan

Part 2 Template

感谢信应态度诚恳、热情洋溢，不能过于简短，以免使收信人有敷衍了事、言不由衷的错觉。

感谢信通常包括以下内容：

一是对对方的馈赠、帮助、关心和邀请等表示由衷地感谢。

二是说明对方的馈赠、帮助、关心和邀请等所起的作用和对自己的意义。

三是表示希望日后有机会答谢对方，并在结尾再次致谢。

Dear ＿＿＿＿＿＿,

　　I am writing to extend my sincere gratitude for ＿＿＿＿＿＿（感谢的原因）. If it had not been for your assistance in ＿＿＿＿＿＿（对方给你的具体帮助）, I fear that I would have been ＿＿＿＿＿＿（没有对方帮助时的后果）. Every one agrees that it was you who ＿＿＿＿＿＿（给出细节）. Again, I would like to express my warm thanks to you! Please accept my gratitude.

Yours sincerely,

＿＿＿＿＿＿

Part 3　Useful Patterns

1. I am writing to express my thanks for...

2. I would like to convey in this letter my heartfelt thanks to you for...

3. Thank you so much for the gift you sent me. It's one of the most wonderful gifts I have ever got.

4. Many thanks for all the good things you have done in helping us to...

5. On behalf of my whole family, I would like to extend my sincere gratitude for...

6. Nothing will be able to erase our wonderful memories, and we will nurture them forever.

7. Thanks again for your kind help.

8. Please accept my thanks.

9. Your help is very much appreciated by each one of us.

10. My genuine gratitude is beyond any words' description.

1. 我写这封信是想表达我对…的谢意。

2. 在信中我要为…表达我真挚的谢意。

3. 非常感谢你给我寄来的礼物，这是我收到的最好礼物之一。

4. 非常感谢您为帮助我们…所做的一切。

5. 我谨代表我全家对…表示真诚的感激。

6. 没有什么能抹掉我们美好的记忆，我们会永远铭刻在心。

7. 对你的帮助再次表示感谢。

8. 请接受我的谢意。

9. 我们每个人都非常感谢你的帮助。

10. 任何语言都不能表达我的真挚谢意。

Part 4 | Exercises

Ⅰ. Translate the following letter of thanks into Chinese.

Dear Mr. Smith,

I am writing to you just to tell you how much I appreciate the warm welcome you extended to my colleague Ms. Jenny Wilson when she visited your company last week. The help and advice you gave her and the introductions you arranged for her, have resulted in a number of very useful meetings. I would like you to know how grateful I am for all you have done to make them possible.

I realized the value of time to a busy person like you, and this makes me all the more appreciative of the time you so generously gave her.

Yours sincerely,

Bill Johnson

Ⅱ. Write a letter of thanks according to the information given in Chinese.

以 Barbara 和 Richard 的名义写一封感谢信给 Dave 和 Jane，表示周六晚参加他们组织的晚宴感到十分满意，而且很高兴认识了他们的朋友 Lucy 和 Kate，觉得他们所讲述的中国假期很有趣，希望将来还有机会再见面。

Unit 4

Computer Security

Unit Goals

After learning this unit, you will be able to:

- talk about computer security
- understand the passages and grasp the key words and expressions
- review the grammar of tense
- write a letter of invitation

Section Ⅰ Listening & Speaking

hard drive 硬盘驱动器	cyber security 网络安全
network sharing 网络共享	network security protection measures 网络安全防
mobile disk 移动硬盘	护措施
anti-virus software 防病毒软件	campus network security 校园网络安全
phishing e-mail 钓鱼邮件	personal information leakage 个人信息泄露
Trojan horse program 特洛伊木马程序	property loss 财产损失
overseas hacker groups and criminals 海外黑客团	firewall 防火墙
体和犯罪分子	software vulnerability 软件漏洞
instant messenger 即时通讯	unknown file 不明文件
data leakage 数据泄露	

Read the conversations carefully and then complete the communicative tasks.

Conversation 1

A：Son, can you tell what's wrong with my computer? I can't move the mouse, I can't use the keyboard. Nothing is working right.

B：Hmm. It looks like your hard drive **crashed**. You just need to **reboot** your computer.

（*a few minutes after Momrebooted the computer*）

A：Oh, my! The computer crashed again! This is the third time today!

B：Oops, your computer may be **infected** with a virus.

A：What is a computer virus?

B：Computer virus is a **program** that threatens computer security. It will destroy the data in the computer and affect the normal operation of the computer.

A: How could my computer get infected with a virus?

B: There are many ways to spread viruses on computers, such as e-mail, network sharing, web browsing, mobile disk transmission, opening files with viruses, and so on.

A: What should we do now?

B: It's very simple, just remove the virus from your computer. By the way, is there any anti-virus software in your computer?

A: I don't know. What is anti-virus software?

B: Anti-virus software is a program tool, which can **detect** whether there is a virus in the computer and remove the detected virus.

A: I've never used it.

B: It doesn't matter. Let me **install** an anti-virus software for you.

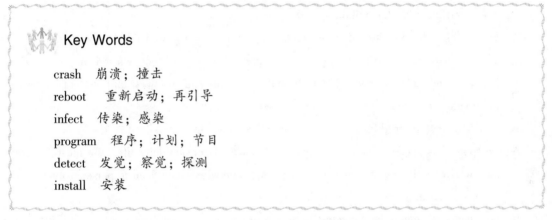

Key Words

crash   崩溃；撞击

reboot   重新启动；再引导

infect   传染；感染

program   程序；计划；节目

detect   发觉；察觉；探测

install   安装

Task 1   Act out the conversation with your partner based on the following clues.

A: Son, can you tell what's wrong with my computer? I can't move the mouse, I can't use the keyboard. Nothing is working right.

B: Hmm. It looks like your hard drive crashed. _____①_____.

(*a few minutes after Momrebooted the computer*)

A: Oh, my! The computer crashed again! This is the third time today!

B: Oops, _____②_____.

A: What is a computer virus?

B: _____③_____. It will destroy the data in the computer and affect the normal operation of the computer.

A: How could my computer get infected with a virus?

B: _____④_____.

A: What should we do now?

B: It's very simple, just remove the virus from your computer. By the way, is there any anti-virus software in your computer?

A: I don't know. What is anti-virus software?

B: _____⑤_____ .

A: I've never used it.

B: It doesn't matter. Let me install an anti-virus software for you.

Conversation 2

A: Xiao Li, have you watched the news yesterday? Northwestern Polytechnical University's e-mail system was **hacked**.

B: No, I haven't. What happened?

A: Overseas hacker groups and criminals sent a batch of phishing emails containing Trojan horse programs to the e-mail system of Northwestern Polytechnical University in an attempt to steal relevant e-mail data and citizens' personal information.

B: It was terrible. Did the **villains** get their way?

A: Fortunately, Northwestern Polytechnical University has always valued cyber security and adopted network security protection measures, which has not caused important data leakage, and campus network security and the security of teachers' and students' personal information have been effectively maintained.

B: I can't believe these things are happening all around us.

A: Computer has become one of the most important tools in our daily life. We must **enhance** the awareness of computer security protection, otherwise it may lead to personal information leakage, and even cause property losses.

B: Will home computers be **targeted** by hackers?

A: Home computers are not worthy of being hacked, but there is still a risk.

B: What cyber security precautions should we take?

A: We need to install firewalls and anti-virus software, and upgrade them regularly to defend against most hackers; and **patch** the system frequently to plug software vulnerabilities. In addition, do not visit unfamiliar websites and open unknown files, etc.

B: Why should we update patches regularly?

A: It is impossible to write a perfect program, so there will be defects inevitably. Regular updates to the latest security patches can effectively prevent illegal **invasion**.

B: Why not open unknown pages, e-mail links or attachments?

A: A large amount of viruses and Trojan horses are likely to be hidden in pages, e-mail links and attachments of unknown origin. Once opened, these viruses and horses will enter the computer and hide in the computer, which will lead to file loss and damage, information leakage and even system crash.

B: The Internet has brought great convenience to our life, but the virus, Trojan horses, hackers make me have a love-hate relationship with it!

 **Key Words**

hack　非法侵入（他人计算机系统）

villain　坏人；恶棍

enhance　提高；增加；加强

target　把…作为目标；瞄准

patch　修补；打补丁　*n.* 补丁

invasion　侵略；侵入

Task 2　Act out the conversation with your partner based on the following clues.

A: Xiao Li, have you watched the news yesterday? Northwestern Polytechnical University's e-mail system was hacked.

B: No, I haven't. What happened?

A: Overseas hacker groups and criminals sent a batch of phishing emails containing Trojan horse programs to the e-mail system of Northwestern Polytechnical University in an attempt to steal relevant e-mail data and citizens' personal information.

B: It was terrible. Did the villains get their way?

A: Fortunately, Northwestern Polytechnical University has always valued cyber security and adopted network security protection measures, which has not caused important data leakage, and campus network security and the security of teachers' and students' personal information have been effectively maintained.

B: I can't believe these things are happening all around us.

A: Computer has become one of the most important tools in our daily life. ____①____.

B: Will home computers be targeted by hackers?

A: ____②____.

B: What cyber security precautions should we take?

A: ____③____; and patch the system frequently to plug software vulnerabilities. In addition, do not visit unfamiliar websites and open unknown files, etc.

B: Why should we update patches regularly?

A: It is impossible to write a perfect program, so there will be defects inevitably. ____④____.

B: Why not open unknown pages, e-mail links or attachments?

A: ____⑤____. Once opened, these viruses and horses will enter the computer and hide in the computer, which will lead to file loss and damage, information leakage and even system crash.

B: The Internet has brought great convenience to our life, but the virus, Trojan horses, hackers make me have a love-hate relationship with it!

Conversation 3

A：Xiaohong! What the…! Ray is sending a lot of porn ads to me. This is **weird**.

B：Oh, my god. May is doing the same thing to me! What happened to them?

A：I bet that their QQ accounts have been stolen.

(*A moment later…*)

B：Xiaohua, I just can't **log in** my QQ account! Report to Tencent!

A：What the hell is going on here? Wow, I just wake up to find that many users of Tencent's QQ **instant** messenger accounts have been hacked to spread porn ads.

B：Spreading porn contents is **illegal** and a severe offense could lead to prison.

A：But what did these criminals spread porn ads for?

B：For money, of course. Just imagine they are in a **shabby** house with a laptop connected with Wi-Fi, working from home and earning money from us.

A：It's funny but it's a shame! They must be in jail.

Key Words

weird　离奇的；古怪的；怪异的

log in　注册

instant　立即的；即时的；速成的

illegal　非法的；不合法的

shabby　破旧的；寒酸的；低劣的

Task 3　Act out the conversation with your partner based on the following clues.

A：Xiaohong! What the…! Ray is sending a lot of porn ads to me. This is weird.

B：Oh, my god. _____①_____ ! What happened to them?

A：_____②_____ .

(*A moment later…*)

B：Xiaohua, _____③_____ ! Report to Tencent!

A：What the hell is going on here? Wow, I just wake up to find that many users of Tencent's QQ instant messenger accounts have been hacked to spread porn ads.

B：_____④_____ .

A：But what did these criminals spread porn ads for?

B：For money, of course. Just imagine they are in a shabby house with a laptop connected with Wifi, working from home and earning money from us.

A：_____⑤_____ ! They must be in jail.

听力材料

Part C   Passages

Listen to the following passages carefully and fill in the blanks with the information you've heard.

Passage 1

Here are a few ideas for _____ strong passwords and keeping them safe:

Use at least 10 characters; 12 is _____ for most home users.

Try to be _____—don't use names, dates, or common words. Mix numbers, symbols, and capital letters into the middle of your password, not at the beginning or end.

Don't use the same password for many _____. If it's stolen from you-or from one of the companies where you do business-thieves can use it to take over all your accounts.

Don't share passwords on the phone, in texts or by e-mail. Legitimate companies will not ask you for your password.

If you write down a password, keep it _____, out of plain sight.

Passage 2

Keep your equipment up-to-date by installing the latest security _____ for your computer-obviously. Less obvious perhaps, is doing so for other technology also _____ to the internet-such as your _____. A recent study carried out by the American Consumer Institute Centre for Citizen Research found that "five out of six routers are inadequately updated for known security _____." These require software updates, too. And don't forget your other _____ devices, such as your smart TV!

Passage 3

A Trojan horse may actually appear to be a useful _____, which is why so many unsuspecting people _____ them. It might be disguised as a _____ intended to rid your computer of viruses, yet actually be used to _____ your system instead. While the terms "virus" and "Trojan horse" are frequently used interchangeably, they are actually quite different. A virus _____ itself, while a Trojan horse does not.

Passage 4

Your Social Security number, _____ card numbers, and bank and utility account numbers can be used to _____ your money or open new accounts in your name. So every time you are asked for your _____ information-whether in a web form, an e-mail, a text, or a phone _____-think about why someone needs it and whether you can really _____ the request.

Passage 5

If you are unsure, test drive a trial version of a _____ or download a free one to check it out. Run it for one day and examine the _____. You'll be shocked by the amount of _____ that goes in and out of your computer in twenty-four hours. You may even detect a Trojan or other _____ that got past your anti-virus, since they are frequently difficult to _____.

Section Ⅱ   Reading

Computer Security

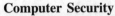

Scammers, **hackers** and **identity** thieves are looking to steal your personal information and your money. But there are steps you can take to protect yourself, like keeping your computer software up-to-date and giving out your personal information only when you have a good reason.

Update Your Software

Keep your software-including your operating system, the web **browsers** you use to connect to the Internet, and your apps-up to date to protect against the latest threats. ① Most software can update automatically, so make sure to set yours to do so.

Outdated software is easier for criminals to break into. If you think you have a virus or bad software on your computer, check out how to detect and get rid of **malware**.

Reasons to Use the Steps of the Scientific Method

Don't hand it out to anyone. Your Social Security number, credit card numbers, and bank and utility account numbers can be used to steal your money or open new accounts in your name. ② So every time you are asked for your personal information-whether in a web form, an e-mail, a text, or a phone message-think about why someone needs it and whether you can really trust the request.

In an effort to steal your information, scammers will do everything they can to appear trustworthy. Learn more about scammers who **phish** for your personal information.

Protect Your Passwords

Here are a few ideas for creating strong passwords and keeping them safe:

Use at least 10 characters; 12 is ideal for most home users.

Try to be **unpredictable**-don't use names, dates, or common words. Mix numbers, symbols, and capital letters into the middle of your password, not at the beginning or end.

Don't use the same password for many accounts. If it's stolen from you-or from one of the companies where you do business-thieves can use it to take over all your accounts. [3]

Don't share passwords on the phone, in texts or by e-mail. **Legitimate** companies will not ask you for your password.

If you write down a password, keep it locked up, out of plain sight.

Back Up Your Files

No system is completely secure. Copy your files to an external hard drive or cloud storage. If your computer is attacked by malware, you'll still have access to your files. [4]

 New Words and Expressions

scammer ['skæmər]   *n.* 诈骗犯

hacker ['hækə(r)]　*n.* 黑客；电脑高手

identity [aɪ'dentəti]　*n.* 身份；特征

update [ˌʌp'deɪt]　*v.* 更新；升级　*n.* 更新

browser ['braʊzə(r)]　*n.* [计算机]浏览器

outdated [ˌaʊt'deɪtɪd]　*adj.* 旧式的；落伍的；过时的

malware ['mælweə(r)]　*n.* 恶意软件

utility [juː'tɪləti]　*n.* 公用事业

phish [fɪʃ]　*v.* 网络钓鱼

unpredictable [ˌʌnprɪ'dɪktəbl]　*adj.* 不可预知的

legitimate [lɪ'dʒɪtɪmət]　*adj.* 合法的；正当的；合理的

Notes

1. Keep your software-including your operating system, the web browsers you use to connect to the Internet, and your apps-up to date to protect against the latest threats.

   Analysis："operating system" 意为 "操作系统"，"web browsers" 意为 "网页浏览器"，"apps" 是 applications 的缩写。破折号中间的部分用来解释说明 software 所包含的具体内容。

   Translation：保持你的软件更新——包括操作系统、用来上网的网页浏览器和应用程序，以防止最新的计算机安全威胁。

2. Your Social Security number, credit card numbers, and bank and utility account numbers can be used to steal your money or open new accounts in your name.

   Analysis："Social Security number" 意为 "社保号码"，"utility account numbers" 意为 "公共服务账户号码"。

   Translation：你的社会保险号码、信用卡号码、银行和公共服务账户号码都可以用来偷你的钱或以你的名义开立新账户。

3. If it's stolen from you-or from one of the companies where you do business-thieves can use it to take over all your accounts.

   Analysis：破折号中间的部分用来对 you 进行补充说明，这里 where 用来引导定语从句，用来修饰 the companies。"take over" 意为 "接管；继承；接收"，这里指网络小偷盗取账户的行为。

   Translation：如果小偷从你处或者某间与你有业务往来的公司窃取了密码，就可以用它来破解你所有的账户。

4. No system is completely secure. Copy your files to an external hard drive or cloud storage. If your computer is attacked by malware, you'll still have access to your files.

   Analysis："external hard drive" 意为 "外部硬盘驱动器"，"cloud storage" 意为 "云存储"。

   Translation：没有任何一个系统是完全安全的。将文件复制到外部硬盘驱动器或云

存储中。如果你的电脑被恶意软件攻击，你仍然可以访问你的文件。

Exercises

Ⅰ. Answer the following questions according to the text.

1. What software should be updated to prevent the latest computer security threats?

_____

_____

2. What is the disadvantage of using outdated software?

_____

_____

3. Why can't hand your personal information out to anyone?

_____

_____

4. What's the suggestion of creating strong passwords?

_____

_____

5. Why do you need to backup files?

_____

_____

Ⅱ. Fill in the blanks with words according to the meaning of the article by memory.

Don't hand it out to anyone. Your Social _____ number, credit card numbers, and bank and utility account numbers can be used to steal your money or _____ new accounts in your name. So every time you are asked for your personal _____ -whether in a web form, an e-mail, a text, or a phone _____ -think about why someone needs it and whether you can really trust the _____. In an _____ to steal your information, scammers will do everything they can to appear _____. Learn more about scammers who _____ for your personal information.

Ⅲ. Fill in the blanks with the words given below. Change the forms when necessary.

thief	automatic	get rid of
easy	use	predict
lock	complete	

1. No system is _____ secure.

2. Your Social Security number, credit card numbers, and bank and utility account numbers

can _____ to steal your money or open new accounts in your name.

    3. If you think you have a virus or bad software on your computer, check out how to detect and _____ malware.

    4. Scammers, hackers and identity _____ are looking to steal your personal information and your money.

    5. Most software can update _____, so make sure to set yours to do so.

    6. Outdated software is much _____ for criminals to break into.

    7. Try to be _____-don't use names, dates, or common words.

    8. If you write down a password, keep it _____ up, out of plain sight.

IV. Translate the following sentences into Chinese.

1. Keep your software-including your operating system, the web browsers you use to connect to the Internet, and your apps-up to date to protect against the latest threats.

_____

_____

2. If you think you have a virus or bad software on your computer, check out how to detect and get rid of malware.

_____

_____

3. Your Social Security number, credit card numbers, and bank and utility account numbers can be used to steal your money or open new accounts in your name.

_____

_____

4. In an effort to steal your information, scammers will do everything they can to appear trustworthy.

_____

_____

5. If it's stolen from you-or from one of the companies where you do business-thieves can use it to take over all your accounts.

_____

_____

Passage 2

Cyber Security and You

It's a dangerous world out there in **cyberspace**. Hackers, **viruses**, and malware exist and are very real threats in cyberspace. A 2017 UK government **cyber-security breaches** survey showed that in the preceding 12 months, just under half of the companies which were asked had identified an attack or breach. [①] That could have been disastrous.

The problem is not about the computers, **firewalls** or **encryption**—it's about us. People are bad at following good cyber-security **protocol**—such as using an effective password. We're also pretty dumb when it comes to clicking on links and downloading content we shouldn't, Ian Pratt, co-founder of cyber-security firm Bromium told the BBC. With that in mind, here are three tips to keep you "cyber-safe".

Passwords are a weakness. The ideal password should be at least eight **characters** long, and contain upper and lower case letters, symbols and numbers. And don't use a common word such as "password". "This makes you **vulnerable** to a **scrape attack**. This is where hackers take the most common passwords and try them on millions of accounts," Thomas Pedersen from OneLogin, an identity and access management company, told the BBC.

Keep your equipment **up-to-date** by installing the latest security updates for your computer-obviously. Less obvious perhaps, is doing so for other technology also connected to the Internet-such as your **router**. [2] A recent study carried out by the American Consumer Institute Centre for Citizen Research found that "five out of six routers are inadequately updated for known security flaws." They require software updates, too. And don't forget your other smart devices, such as your smart TV!

Finally, consider enabling multi-factor **authentication** for your accounts. That's where a mobile phone or **dongle** is used to verify access to a device. If your password becomes compromised, no access can be obtained without your second authentication device… so don't lose your phone! [3] In the future, **biometrics** such as voice and fingerprint may make this easier.

The Internet is wonderful, but security threats do exist. Fortunately, with a little common sense and forethought, you can be secure from the majority of cyber-security threats.

New Words and Expressions

cyberspace ['saɪbəspeɪs]　*n.* 网络空间

virus ['vaɪrəs]　*n.* 计算机病毒

cyber-security ['saɪbə sɪ'kjʊərəti]　*adj.* 网络安全的

breach [briːtʃ]　*n.* 破坏，违规

firewall ['faɪəwɔːl]　*n.* 防火墙

encryption [ɪn'krɪpʃn]　*n.* 数据加密

protocol ['prəʊtəkɒl]   *n.* 协议、协定

character ['kærəktə(r)]   *n.* 字符(如数字、字母、标点等)

vulnerable ['vʌlnərəbl]   *adj.* 易受攻击的

scrape attack   撒网式密码盗窃

up-to-date   最新的

router ['ruːtə(r)]   *n.* 路由器

flaw [flɔː]   *n.* 漏洞

multi-factor authentication   多重身份认证

dongle ['dɒŋgl]   *n.* 软件保护器;加密狗

compromised ['kɒmprəmaɪz]   *adj.* 被攻破的

biometrics [ˌbaɪəʊ'metrɪks]   *n.* 生物特征识别技术;生物测定学

Notes

1. A 2017 UK government cyber-security breaches survey showed that in the preceding 12 months, just under half of the companies which were asked had identified an attack or breach.

   Analysis:"IDENTIFIED" 意为 "识别,辨明"。

   Translation:2017 年,英国政府的一项网络安全漏洞调查显示,在过去 12 个月里,被询问的公司中只有不到一半发生了攻击或入侵。

2. Keep your equipment up-to-date by installing the latest security updates for your computer-obviously. Less obvious perhaps, is doing so for other technology also connected to the Internet-such as your router.

   Translation:通过安装最新的安全更新可以保持您的设备的更新——这是显而易见的。或许不那么明显的是要对其他连接到互联网的技术也这样做——比如你的路由器。

3. If your password becomes compromised, no access can be obtained without your second authentication device…so don't lose your phone!

   Analysis:"access" 意为 "访问","authentication" 意为 "身份验证"。

   Translation:如果你的密码被泄露了,没有第二个身份验证设备就无法访问,所以不要丢失你的手机!

Exercises

Answer the following questions according to the text.

1. Why is cyberspace a dangerous world?

_____

_____

2. What is the real cause of cyber security problems?

_____

_____

3. What is the ideal password?

_____

_____

4. Why not use common words as a password?

_____

_____

5. Why is it recommended to enable multi-factor authentication for your accounts?

_____

_____

Passage 3

What is a Trojan Horse?

　　Named for a giant horse that was supposed to be a gift but was filled with the Greek army, a Trojan horse program can be just as **deceptive**. [1] The story goes that the Greeks gave the Trojans a huge wooden horse as a peace offering. The citizens of Troy accepted the gift, brought the horse inside the city, threw a victory **bash**, then went to bed. The Trojans didn't realize they'd been taken until the Greek soldiers had set the city on fire. A Trojan horse that affects computers can contain some **nasty** surprises as well. It can damage, delete, or destroy important files.

A Trojan horse may actually appear to be a useful application, and that's why so many **unsuspecting** people download them. It might be **disguised** as a program intended to get rid of your computer viruses, yet actually it is used to infect your system instead. While the terms "virus" and "Trojan horse" are frequently used **interchangeably**, they are actually quite different. A virus **replicates** itself, while a Trojan horse does not.

Once a Trojan horse is **activated**, it can access files, **folders**, or your entire system. Commonly, Trojans create a "backdoor" or a "trapdoor", which can be used to send your personal information to another location. To protect your system, extensive anti-virus software is a good first step. Choose a program that looks for Trojan horses and **worms** as well as viruses, and make sure it updates **definitions** for each frequently. Also, make sure your anti-virus scans e-mail, and gives an alert or automatically deletes any message that contains **suspicious** code, even if the code is not specific malware that the program has already identified. [2]

The most effective option for protecting against a Trojan horse may be installing a firewall if you do not already have one. A good software firewall is usually the best type for a personal computer. It can be **configured** to keep **intruders** out and keep your system, or programs within your system, from sending out personal or confidential data. [3]

If you are unsure, test drive a trial version of a firewall or download a free one to check it out. Run it for one day and examine the log file. You'll be shocked by the amount of information that goes in and out of your computer in twenty-four hours. You may even detect a Trojan or other malware that got past your anti-virus, since they are frequently difficult to detect.

New Words and Expressions

deceptive [dɪˈseptɪv]   *adj.* 骗人的；虚伪的；诈欺的

bash [bæʃ]   *n.* 狂欢

nasty [ˈnɑːsti]   *adj.* 危害的；严重的；令人不快的

delete [dɪˈliːt]   *v.* 删除

unsuspecting [ˌʌnsəˈspektɪŋ]   *adj.* 无疑心的；无戒心的；未料到的

disguise [dɪsˈɡaɪz]   *v.* 掩饰；假装；假扮

interchangeably [ˌɪntəˈtʃeɪndʒəblɪ]   *adv.* 可交换地

replicate [ˈreplɪkeɪt]   *v.* 复制；模拟

activate [ˈæktɪveɪt]   *v.* 激活；使活动；起动

folder [ˈfəʊldə(r)]   *n.* 文件夹

extensive [ɪkˈstensɪv]   *adj.* 广泛的；广阔的；大量的

worm [wɜːm]   *n.* 蠕虫

definition [ˌdefɪˈnɪʃn]   *n.* 定义；阐释；清晰度

suspicious [səˈspɪʃəs]   *adj.* 可疑的；多疑的

configure [kənˈfɪɡə]   *v.* 配置

intruder [ɪnˈtruːdə(r)]   *n.* 入侵者；干扰者；妨碍者

 Notes

1. Named for a giant horse that was supposed to be a gift but was filled with the Greek army, a Trojan horse program can be just as deceptive.

   Analysis："name" 在这里是动词，意为"给…取名"。"be supposed to be" 意为"被认为是…"。

   Translation：特洛伊木马程序取名自希腊神话特洛伊木马记。一匹本应作为礼物的巨马，但里面装满了希腊军队，木马程序同样具有欺骗性。

2. Also, make sure your anti-virus scans e-mail, and gives an alert or automatically deletes any message that contains suspicious code, even if the code is not specific malware that the program has already identified.

   Analysis：此句是祈使句，make sure 后面的部分是宾语从句。这个宾语从句由三个并列句组成，主语都是"anti-virus"，意为"反病毒软件"，谓语分别是 scans, gives 和 deletes。

   Translation：此外，确保你的反病毒软件会扫描电子邮件，并发出警报或自动删除任何包含可疑代码的消息，即使这些代码不是程序已经识别的特定恶意软件。

3. It can be configured to keep intruders out while also keeping your system, or programs within your system, from sending out personal or confidential data.

   Translation：它可以被配置为防止入侵者进入，同时也可以防止你的系统或系统内的

程序发送个人数据或机密数据。

 Exercises

Translate the following short passages into Chinese.

1. The citizens of Troy accepted the gift, brought the horse inside the city, threw a victory bash, then went to bed. The Trojans didn't realize they'd been taken until the Greek soldiers had set the city on fire.

_____

_____

_____

2. A Trojan horse may actually appear to be a useful application, and that's why so many unsuspecting people download them. It might be disguised as a program intended to get rid of your computer viruses, yet actually it is used to infect your system instead.

_____

_____

_____

3. Once a Trojan horse is activated, it can access files, folders, or your entire system. Commonly, Trojans create a "backdoor" or a "trapdoor", which can be used to send your personal information to another location.

_____

_____

_____

Section Ⅲ　Grammar

Tense（时态）

在英语语法中，"时"指动作发生的时间，"态"指动作的状态和性质。这就是英语动词的时态。从时间上看，英语的时态有现在、过去、将来、过去将来之分。从方式上看，英语的时态又有一般、进行、完成、完成进行之分。动词的动作可发生于 4 种不同的时间，表现在 4 种不同的方面。英语的时态是一种动词形式，不同的时态用以表示不同的时间与方式。所以英语动词共有 16 种时态。

一、现在以动词 work 为例，将 16 种不同的时态列表：

时间 形态	现在时	过去时	将来时	过去将来时
一般时态	①一般现在时 I work.	②一般过去时 I worked.	③一般将来时 I will/shall work.	④一般过去将来时 I would/should work.
进行时态	⑤现在进行时 I am working.	⑥过去进行时 I was working.	⑦将来进行时 I will/shall be working.	⑧过去将来进行时 I would/should be working.
完成时态	⑨现在完成时 I have worked.	⑩过去完成时 I had worked.	⑪将来完成时 I will/shall have worked.	⑫过去将来完成时 I would/should have worked.
完成进行 时态	⑬现在完成进行时 I have been working.	⑭过去完成进行时 I had been working.	⑮将来完成进行时 I will/shall have been working.	⑯过去完成进行时 I would/should have been working.

二、下面重点讲解动词的几个较难理解的时态形式。

（一）过去完成时

构成：过去完成时由"助动词 had + 过去分词"构成，其中 had 通用于各种人称。

They had already had breakfast before they arrived at the hotel.

He had finished writing the composition by 10 o'clock this morning.

过去完成时的判断依据

1. 由时间状语来判定。

一般说来，各种时态都有特定的时间状语。与过去完成时连用的时间状语有：

（1）by + 过去的时间点。

I had finished reading the novel by nine o'clock last night.

（2）by the end of + 过去的时间点。

We had learned over three thousand English words by the end of last term.

（3）before + 过去的时间点。

We had planted seven hundred trees before last Wednesday.

2. 由"过去的过去"来判定。

过去完成时表示"过去的过去",是指过去某一动作之前已经发生或完成的动作,即动作有先后关系,动作在前的用过去完成时,在后的用一般过去时。这种用法常出现在:

(1) 宾语从句中。

当宾语从句的主句为一般过去时,且从句的动作先于主句的动作时,从句要用过去完成时。在 told, said, knew, heard, thought 等动词后的宾语从句。

She said that she had seen the film before.

(2) 状语从句中。

在时间、条件、原因、方式等状语从句中,主、从句的动作发生有先后关系,动作在前的,要用过去完成时,动作在后的要用一般过去时。

When I got to the station, the train had already left.

After he had finished his homework, he went to bed.

(3) 表示意向的动词,如 hope, wish, expect, think, intend, mean, suppose 等,用过去完成时表示"原本…,未能…"。

We had hoped that you would come, but you didn't.

3. 根据上下文来判定。

I met Wang Tao in the street yesterday. We hadn't seen each other since he went to Beijing.

过去完成时的主要用法:

1. 过去完成时表示一个动作或状态在过去某一时间或动作之前已经完成或结束,即发生在"过去的过去"。

When I woke up, it had stopped raining.

我醒来时,雨已经停了。(主句的动作发生在"过去的过去")

2. 过去完成时是一个相对的时态,表示的是"过去的过去",只有和过去某一时间或某一动作相比较时才使用它。

He told me that he had written a new book. (had written 发生在 told 之前)

3. 过去完成时需要与一个表示过去的时间状语连用,它不能离开过去时间而独立存在。此时多与 already, yet, still, just, before, never 等时间副词及 by, before, until 等引导的短语或从句连用。

Before she came to China, Grace had taught English in a middle school for about five years.

Peter had collected more than 300 Chinese stamps by the time he was ten.

4. 过去完成时表示某一动作或状态在过去某时之前已经开始,一直延续到这一过去时间,而且动作尚未结束,仍然有继续下去的可能。

By the end of last year, he had worked in the factory for twenty years.

(had worked 已有了 20 年,还有继续进行下去的可能)

5. 过去完成时也用于 hardly…when… (刚…就…), no sooner…than… (刚…就…), It was the first time + that 等一些固定句型中。

He had no sooner left the room than they began to talk about him. 他刚离开房间,他们就议

论起他来。

We had hardly begun when we were told to stop. 我们刚开始就被叫停。

It was the first time that he had ever spoken to me in such a tune. 他用这样的语调跟我讲话，这是第一次。

过去完成时与现在完成时的区别

现在完成时表示的动作发生在过去，但侧重对现在产生的结果或造成的影响，与现在有关，其结构为"助动词 have（has）＋过去分词"；过去完成时则是一个相对的时态，它所表示的动作不仅发生在过去，更强调"过去的过去"，只有和过去某时或某动作相比较时，才用到它。

I have learned 1,000 English words so far. 到目前为止，我已经学会了 1 000 个英语单词。

I had learned 1,000 English words till then. 到那时为止，我已经学会了 1 000 个英语单词。

— I'm sorry to keep you waiting. 对不起，让你久等了。

— Oh, not at all. I have been here only a few minutes.

没什么，我只等了几分钟。（"等"的动作从过去某一时间点持续到现在）

— John returned home yesterday. 约翰昨天回到家的。

— Where had he been? 他去哪儿了？

（答语中使用过去完成时是指约翰在 returned home 之前去了哪些地方，即"过去的过去"）

（二）将来完成时

构成：shall/will have done sth.

用法：

1. 表示在将来某一时刻或某一时刻之前已完成的动作，并往往对将来某一时间产生影响，常与表示将来的时间状语及条件或时间状语从句连用。

How many words shall we have learned by the end of this term?

到本学期末我们学了多少词汇？

Next Monday I shall have been for a month. 到下周一，我到这就满一个月了。

If you come at five o'clock, I shall not yet have finished dinner. 你若 5 点来，我还没吃完晚饭。

When you get home, she will have gone to bed. 你到家时，她该已经睡了。

2. 表示一种推测。主要用于第二、第三人称。

She will have watched this film already. 她恐怕已经看过这场电影了。

You will have arrived home by now. 这时候你可能已经到家了。

下表为常与完成时连用的句型和时间状语：

句型/时间状语	完成时态
It is/will be the first time（that）	现在完成时
It is the first + sth.（that）	现在完成时

续表

句型/时间状语	完成时态
It is the + 形容词最高级 + sth.（that）	现在完成时
by now；up till now；since + 过去时间（的句子）	现在完成时
It was the first time（that）	过去完成时
by the end of + 过去时间	过去完成时
by the end of + 将来时间	将来完成时
by this time + 将来时间	将来完成时
by the time/when + 一般现在时的从句	将来完成时

（三）现在完成进行时

构成：have/has been + 现在分词

用法：用于表示动作从过去某一时间开始一直延续到现在或离现在不远的时间，动作是否继续下去，视情况而定。

How long has it been raining? 雨下多久了？

She has been sitting there for more than 2 hours. 她已经在那坐了两个多小时了。

We've been seeing quite a lot of each other recently. 最近我们常常见面。

He has been telephoning me several times in two days. 这两天他打了好几次电话给我。

Exercises

Choose the best answer.

1. We are confident that we _____ our export sales by 15 percent by the end of this year.

    A. increase　　　B. had increased　　C. would increase　　　D. will have increased

2. In the last few years, our company _____ a great deal of attention to build up company culture.

    A. pays　　　　B. would pay　　　C. is paying　　　D. has be paying

3. As soon as we _____ your order, we will process it and deliver your package immediately.

    A. receive　　　B. had received　　C. received　　　D. will receive

4. We are sure that we _____ our second project by the end of the next month.

    A. will have completed　　　　B. had completed

    C. completed　　　　　　　　D. were completing

5. By the end of next year, I _____ for the company for 10 years.

    A. work　　　　B. am working　　C. had worked　　D. will have worked

6. According to the time table, the train for Beijing _____ at 9：10 p. m. from Monday to Friday.

    A. was leaving　　B. is leaving　　C. leaves　　　D. has left

7. They will not start the project until the board chairman _____ back from South Africa.

   A. will come   B. is coming   C. came   D. comes

8. She was quite sure that she _____ the door before she left the office.

   A. will lock   B. would lock   C. has locked   D. had locked

9. We don't have to hurry as the bus _____ for London at five in the evening.

   A. leaves   B. left   C. has been leaving   D. has left

10. Last week two engineers _____ to solve the technical problems of the project.

   A. have sent   B. were sent   C. sent   D. had sent

Section IV Writing

Letter of Invitation（邀请信）

邀请信是邀请亲朋好友或知名人士、专家等参加某项活动时所发出的约请性书信。在国际交往以及日常的各种社交活动中，这类书信使用广泛。

邀请信分为两种：一种是正规的格式（formal correspondence），也叫作请柬；一种是非正规格式（informal correspondence），就是一般的邀请信。

Part 1 Sample

The following is a letter of invitation. Please read and try to understand it.

April 5，2021

Dear Prof. Smith，

An international innovation contest will be held at our university from September 24th－29th and we are very glad to invite you to organize a team for the event.

The innovation contest is targeted to talented youth exploring the sea of innovative ideas. And prizes will be awarded against three criteria：innovation degree, professional level, and practical value. Please find more details in the attached file.

You are so welcomed to lead a team in the innovation contest. Please confirm your availability on or before August 1st by replying to this e-mail. Your consideration of this invitation will be highly appreciated！

Sincerely yours，
Liu Ming

Part 2 Template

从上面的样例可以看出邀请信（letter of invitation）写作要领如下：

1. 邀请信一定要将邀请对方参加活动的内容、时间（年、月、日、钟点）、地点、场合写清楚，不能让收信人有任何的疑虑。例如"I'd like you and Tom to come to dinner next Monday"这句话中指的是哪个星期并不明确，所以要加上具体日期，"I'd like you and Tom to come to dinner next Monday，May the fifth at 6 o'clock. p. m. ，at the Locus Hotel"。

2. 邀请信的特点是简短热情，形式较为灵活，叙事要清楚、明白。一般为 2～3 段，将意思表达清楚即可。如写给朋友，可选用活泼、真诚的言辞；如写给长辈、上级、名人等，则语言应正式、礼貌。

3. 在盛大的场合，邀请信应提前三星期发出，一般场合书写邀请信，应在预定时间的前几天发出，以便收信人及早考虑和安排。收到信后，通常应立即答复，表明接受或谢绝，以示礼貌。

Mar. 6，2019 Dear _____， 　　We would like to invite you to _____（1） _____ _____. 　　Everyone is going to catch _____（2） _____ _____. 　　（3）_____ _____ 　　　　　　　I hope you will be able to come. 　　　　　　　　　　　　　Sincerely， 　　　　　　　　　　　　　Mark	写信日期 Date 称呼 Salutation to invitee 正文 Body （1）写信目的（事由） （2）具体说明（时间、地点、内容等） （3）表达感激与期望 结束句 Complimentary Close 署名 Signature（inviter）

Part 3 Useful Patterns

1. I'd like you to come to our dinner this evening.	1. 我想让你今晚来参加我们的晚餐。
2. Request the pleasure of…	2. 恭请…
3. We sincerely hope you can attend.	3. 我们真诚地希望你能参加。
4. I hope you are not too busy to come.	4. 期望您在百忙中光临。
5. I'm looking forward to seeing you.	5. 我期待着见到您。
6. We thought it would be pleasant to have some of our friends to celebrate it.	6. 我们认为有一些朋友来庆祝会很愉快。
7. The reception will be held in…, on…	7. 招待会定于…在…举行。
8. We have decided to have a party in honor of the occasion.	8. 为此我们决定举行一次晚会。
9. Please let me know as soon as possible, If you can come and tell me when you will be able to do so.	9. 如能来的话，早日告我，何时为宜。
10. Please confirm your participation at your earliest convenience.	10. 是否参加，请早日告之。

Part 4 Exercises

Ⅰ. Translate the following letter of invitation into Chinese.

Dear Karen and Michael，

　　Thank you so much for including us in the group that will be celebrating your anniversary. Tenyears！ What a wonderful milestone for such a wonderful couple.

　　Unfortunately，we will not be able to attend the party. Our daughter Heather will graduate from high school that day and we'll be spending our time with her.

I will call you soon, though, to see if we can get together for dinner to celebrate our anniversary and raise a toast to our new college student.

Sincerely,

Louise and Greg

II. Write a letter of invitation according the information given below.

Diamond 夫妇邀请 Louise 和 Greg 夫妇于 11 月 30 日星期五晚上八点到家里参加婚礼十周年的聚会，已准备好香槟和蛋糕，让朋友们带来一些制造热闹气氛大的彩带（confetti），不需带礼物。如果有空前来参加，请告知。

Unit 5

Internet

Unit Goals

After learning this unit, you will be able to:

- talk about Internet
- understand the passages and grasp the key words and expressions
- review the grammar of passive voice
- write a letter of complaint

Section I Listening & Speaking

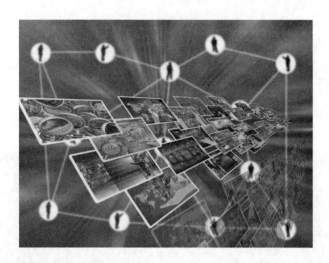

Part A Words and Expressions

browse Tik Tok 刷抖音

Internet plus 互联网 +

information age 信息时代

knowledge society 知识社会

traditional industry 传统工业

information and Internet platforms 信息和网络平台

Online shopping 在线购物

shopping websites 购物网站

big data 大数据

cloud computing 云计算

high-quality educational resources 高质量教育资源

surf the Internet 上网

broadenone's mind 开阔某人的视野

enrichone's knowledge 丰富某人的知识

addict to the Internet 沉迷网络

spare time 闲暇时间

life style 生活方式

community services 社区服务

Part B Conversations

Read the conversations carefully and then complete the communicative tasks.

Conversation 1

A：Try some peaches I just bought. They are delicious.

B：Mm, It tastes good. Where did you buy it?

A：I found a nearby peach orchard when I **browse** TikTok the other day, and I had a picking tour with my family yesterday.

B：Wow, this is really a typical Internet plus **agriculture** and **tourism**!

A：The term "Internet plus" is often heard these days. What is Internet plus?

B："Internet Plus" is a new form of Internet development in China under the innovative form of information age and knowledge society.

A: I still don't understand. Can you explain it more easily?

B: To put it simply, it is "Internet plus traditional industry", which makes use of information and Internet platforms to integrate the Internet with traditional industries, and makes use of the advantages and characteristics of the Internet to create new development opportunities.

A: I get it. Online shopping can also be regarded as "Internet plus retail", right?

B: Yes. Major shopping websites rely on big data, cloud computing and other technologies to master market information and consumption data, meet consumer needs, and realize the combination of Internet and retail.

A: Can "Internet plus" also be achieved in education?

B: In the **context** of openness, high-quality educational resources are integrated through the Internet, enabling people to access the learning resources they want anytime and anywhere.

A: Are there any more examples closely related to our lives?

B: Too **much**. For example, Internet plus tourism, Internet plus **logistics**, Internet plus medical care, Internet plus **community** services, etc. The Internet has played a huge role in various fields of consumption, production and life.

A: The Internet has really brought us a lot of convenience.

B: Yes, I believe that "Internet plus" will bring us more surprises in the future.

Key Words

browse　翻阅，浏览（信息）

agriculture　农业

tourism　旅游业

context　（想法、事件等的）背景；上下文

logistics　物流

community　社会（团体）

Task 1　Act out the conversation with your partner based on the following clues.

A: Try some peaches I just bought. They are delicious.

B: Mm, It tastes good. Where did you buy it?

A: I found a nearby peach orchard ＿＿＿＿①＿＿＿＿, and I had a picking tour with my family yesterday.

B: Wow, ＿＿＿＿②＿＿＿＿ !

A: The term "Internet plus" is often heard these days. What is Internet plus?

B: "Internet Plus" is a new form of Internet development in China under ＿＿＿＿③＿＿＿＿.

A: I still don't understand. Can you explain it more easily?

B: To put it simply, it is "Internet plus traditional industry", which makes use of information and Internet platforms to _____④_____, and makes use of the advantages and characteristics of the Internet to create new development opportunities.

A: I get it. Online shopping can also be regarded as "Internet plus retail", right?

B: Yes. Major shopping websites rely on big data, cloud computing and other technologies to master market information and consumption data, meet consumer needs, and realize the combination of Internet and retail.

A: Can 'Internet plus' also be achieved in education?

B: In the context of openness, high-quality educational resources are integrated through the Internet, _____⑤_____ anytime and anywhere.

A: Are there any more examples closely related to our lives?

B: Too much. For example, Internet plus tourism, Internet plus logistics, Internet plus medical care, Internet plus community services, etc. the Internet has played a huge role in various fields of consumption, production and life.

A: The Internet has really brought us a lot of convenience.

B: Yes, I believe that "Internet plus" will bring us more surprises in the future.

Conversation 2

A: With the development of computer technology, the Internet has become more and more popular.

B: Yeah, but do you think students should **surf** the Internet after class. ?

A: Yes, I do. I **regard** it **as** a great helper.

B: Oh? Why?

A: For example, you can surf the Internet for any information you need in a short time without working hard in the library.

B: Oh, it sounds so convenient.

A: Yeah, it is convenient to communicate with others by using the Internet.

B: Through the network, we can also do many other things, such as reading books, **entertaining**, trading stocks, chatting and so on.

A: Yes, these are all the benefits of the web.

B: However, many people think that there are many disadvantages, because there is also some information that is not good for students.

A: That's also true. Well, it will not only have a bad effect on study but also do harm to our health if we spend too much time to play games online.

B: What should we do?

A: We should make proper use of the Internet. It is of great importance for us to separate good plants from wild **weeds**.

B: You are right. We should be the **masters** of the network, not its **slaves**.

Key Words

surf　冲浪；（互联网上）冲浪

regard as　把…认作；当做；看做

entertain　招待；娱乐

weed　杂草，野草

master　主人；硕士

slave　奴隶

Task 2　Act out the conversation with your partner based on the following clues.

A：With the development of computer technology, the Internet has become more and more popular.

B：Yeah, but do you think students should surf the Internet after class?

A：Yes, I do. _____①_____.

B：Oh? Why?

A：For example, you can surf the Internet for any information you need in a short time without working hard in the library.

B：Oh, it sounds so convenient.

A：Yeah, _____②_____.

B：Through the network, we can also do many other things, such as reading books, entertaining, trading stocks, chatting and so on.

A：Yes, these are all the benefits of the web.

B：However, many people think that there are many disadvantages, because there is also some information that is not good for students.

A：That's also true. Well, _____③_____.

B：What should we do?

A：We should make proper use of the Internet. _____④_____.

B：You are right. _____⑤_____.

Conversation 3

A：You look **sleepy**.

B：Yes. I didn't go to bed until two o'clock this morning.

A：You must chat on the Internet.

B：Yes. How do you know that?

A：Everyone knows that you are crazy about suffering the Internet.

B：Really?

A： In my point of view, it is a waste of time to surf the Internet.

B： Maybe you are right. But it's truly very interesting.

A： That's why you just can't give it up?

B： Partly. Apart from chatting on the net, I can learn a lot of knowledge from the Internet.

A： The Internet can indeed **broaden** our mind and **enrich** our knowledge, but **addicting** to it is harmful. It will not only waste of time but harmful to your health.

B： It seems that I should reduce the time of surfing the Internet.

A： Yes. There are plenty of other things to do.

B： What do you usually do in your spare time?

A： I always do exercise, read books and play with my friends.

B： Your life style is very healthy, and I need to change my way of life.

 Key Words

sleepy　困倦的

broaden　变宽；扩大…的范围

enrich　充实；使丰富

addict　使沉迷，使上瘾

Task 3　Act out the conversation with your partner based on the following clues.

A： _____①_____.

B： Yes. I didn't go to bed until two o'clock this morning.

A： _____②_____.

B： Yes. How do you know that?

A： _____③_____.

B： Really?

A： In my point of view, it is a waste of time to surf the Internet.

B： Maybe you are right. But it's truly very interesting.

A： That's why you just can't give it up?

B： Partly. Apart from chatting on the net, I can learn a lot of knowledge from the Internet.

A： _____④_____. It will not only waste of time but harmful to your health.

B： It seems that I should reduce the time of surfing the internet.

A： Yes. _____⑤_____.

B： What do you usually do in your spare time?

A： I always do exercise, read books and play with my friends.

B： Your life style is very healthy, and I need to change my way of life.

听力材料

Part C   Passages

Listen to the following passages carefully and fill in the blanks with the information you've heard.

Passage 1

Internet is a network where users can _____ other users and services around the world via computers. The long history and lots of use cases make the Internet very _____. The Internet provides a lot of different _____ from web browsing to banking, _____ to e-commerce. A lot of real-world processes are _____ into the Internet.

Passage 2

It provides both individuals and businesses with easy access to government services. It _____ the building of an industrial internet. It _____ agricultural improvements and increased income for farmers through the development of _____ e-commerce. It has also popularized medical service _____ such as telemedicine, online appointment and day surgery, and _____ the sharing of medical information among healthcare institutions.

Passage 3

On September 29, China Internet Network Information Center (CNNIC) _____ the 46th Statistical Report on the Development of China's Internet. The number of Internet users in China had _____ 940 million as of June, accounting for one-fifth of the world's _____, according to the report. The Internet _____ rate in China reached 67 percent, about 5 percentage points higher than the global _____, the report said.

Passage 4

Internet is a _____ network which contains a lot of different protocols, _____, products services etc. So the man who invented internet exact question will be no single man. There are a lot of different _____ to the creation of the Internet. But TCP/IP is the heart of the internet and without TCP/IP _____ Internet will not work. So TCP/IP _____ Vinton Cerf and Bob Kahn are called as the creators of the Internet.

Passage 5

The Internet has also been reaching _____ and people in rural areas, gradually _____ the digital divide among the technologically challenged. Last year, 57.6 percent of the rural population had _____ to the Internet, while 43.2 percent of people age 60 and older used the Internet. Residents in 98 percent of China's poor villages had access to the Internet through fiber-optic cables, said the report, noting that the urban-rural _____ had been significantly narrowed. Instant messaging through WeChat and other apps has been the biggest _____ for people to get online in the past decade.

Section II Reading

Passage 1

Who Invented the Internet?

The Internet is the most popular technology term known by billions of people around the world. What makes the most popular is its usage, access range, abilities, and new opportunities. [①] Internet is the core of everyday life like a government, bank, etc. But who invented the Internet?

What is Internet?

Before answering our questions, we should **define** what the Internet is. Internet is a network where users can **access** other users and services around the world via computers. [②] The long history and lots of use cases make the Internet very complex. The Internet provides a lot of different services from web browsing to banking, from **instant** messaging to e-commerce. A lot of real-world processes are integrated into the Internet.

The History of Internet

Even Internet history provides a lot of details. We will talk about some important points of the Internet history. First workable Internet is created by the US Department of **Defence** with the name of Advanced Research Projects Agency Network simply ARPANET. As the name suggests, ARPANET is created with the aim of research. The computers connected to ARPANET can **communicate** and share resource with each other without any physical location **restriction**. The first message and communication in ARPANET occurred between UCLA and Stanford University. The message was LOGIN but only two letters are received and the ARPANET crashed.

What **transformed** ARPANET in a real global network is the TCP/IP communication model and protocols that are created by Robert Kahn and Vinton Cerf. TCP/IP is created at the end of the 1970s and integrated into the ARPANET in 1983. This turned ARPANET a real worldwide network called Internet.

Internet vs WWW (World Wide Web)

We have learned what the Internet is and how it was invented. But what about the web or world wide web? WWW is not the Internet. World wide web is a group of protocols used to create web sites, web pages, and web applications with web servers and client applications named web browser. [3] We can say that the web is not the Internet but works with the help of the Internet by using the Internet.

Who Invented the Internet?

Internet is a huge network which contains a lot of different protocols, inventions, products services etc, so "who invented the Internet" the answer will be no single man. [4] There are a lot of different **contributors** to the creation of the Internet. While TCP/IP is the heart of the Internet. Without TCP/IP protocol Internet will not work, so TCP/IP creators Vinton Cerf and Robert Kahn are called as the creators of the Internet.

TCP/IP and Internet Inventors Vinton Cerf and Robert Kahn

We can also say Tim Berners Lee the inventor of the Web which made the Internet more and more popular with a huge population.

Sir Tim Berners-Lee who invented the World Wide Web in 1989

 New Words and Expressions

core [kɔː(r)]　*n.* 核心；果心；要点

define [dɪˈfaɪn]　*v.* 定义；解释

access [ˈækses]　*v.* 访问，存取(计算机信息)

instant [ˈɪnstənt]　*adj.* 立即的；即时的；速成的

defence [dɪˈfens]　*n.* 防御

communicate [kəˈmjuːnɪkeɪt]　*v.* 沟通；交流

restriction [rɪˈstrɪkʃn]　*n.* 限制；约束

transform [trænsˈfɔːm]　*v.* 改变；转换

protocol [ˈprəʊtəkɒl]　*n.* 协议

server [ˈsɜːvə(r)]　*n.* 服务器

contributor [kənˈtrɪbjʊtə]　*n.* 贡献者；捐助者

 Notes

1. What makes the most popular is its usage, access range, abilities, and new opportunities.

   Analysis："What makes the most popular" 是 "make + 宾语 + 形容词" 的结构，意为 "使…处于某种状态"。后文的 its 指代的是互联网的，从这里可以看出，make 的宾补结构里省略了宾语 Internet。

   Translation：(令互联网) 最受欢迎的是它的用法、访问范围、功能和新机会。

2. Internet is a network where users can access other users and services around the world via computers.

   Analysis：这句话里包含由 where 引导的定语从句，其先行词是 network。

   Translation：互联网是用户可以通过计算机访问世界各地其他用户和服务的网络。

3. World wide web is a group of protocols used to create web sites, web pages, and web applications with web servers and client applications named web browser.

Analysis："web servers" 意为"网络服务器"，"client applications" 意为"客户端应用程序"。

Translation：万维网是一组协议，用于使用网络服务器和名为网络浏览器的客户端应用程序创建网站、网页和网络应用程序。

4. Internet is a huge network which contains a lot of different protocols, inventions, products services etc, so "who invented the Internet" the answer will be no single man.

Translation：互联网是一个庞大的网络，其中包含许多不同的协议、发明、产品服务等。因此，"谁发明了互联网"，答案不是一个人。

Exercises

Ⅰ. **Answer the following questions according to the text.**

1. What makes the Internet the most popular technology term known by billions of people around the world?

_____

_____

2. What is the Internet?

_____

_____

3. Who created the TCP/IP communication model and protocols?

_____

_____

4. Is there any relationship between WWW and the Internet?

_____

_____

5. Who invented the World Wide Web?

_____

_____

Ⅱ. **Fill in the blanks with words according to the meaning of the article by memory.**

Even Internet history provides a lot of details. We will talk about some important _____ of the Internet history. First _____ Internet is created by the US Department of Defence with the name of _____ Research Projects Agency Network simply ARPANET. As the name _____

ARPANET is created with the _____ of research. The computers connected to ARPANET can communicate and share _____ each other without any physical location _____. The first message and communication in ARPANET occurred between UCLA and Stanford University. The message was LOGIN but only two letters are _____ and the ARPANET crashed.

III. Fill in the blanks with the words given below. Change the forms when necessary.

popular	work	connect
receive	create	contain
contribute	with	

1. There are a lot of different _____ to the creation of the Internet.

2. What transformed ARPANET in a real global network is the _____ of the TCP/IP communication model and protocols that are created by Robert Kahn and Vinton Cerf.

3. The computers _____ to ARPANET can communicate and share resource each other without any physical location restriction.

4. The Internet is _____ technology term known by billions of people around the world.

5. First _____ Internet is created by the US Department of Defence with the name of Advanced Research Projects Agency Network simply ARPANET.

6. The message was LOGIN but only two letters _____ and the ARPANET crashed.

7. Internet is a huge network which _____ a lot of different protocols, inventions, products services etc.

8. But TCP/IP is the heart of the Internet and _____ TCP/IP protocol internet will not work.

IV. Translate the following sentences into Chinese.

1. The Internet is the most popular technology term known by billions of people around the world.

_____

_____

2. Internet is a network where users can access other users and services around the world via computers.

_____

_____

3. The Internet provides a lot of different services from web browsing to banking, instant messaging to e-commerce.

_____

_____

4. The computers connected to ARPANET can communicate and share resource each other without any physical location restriction.

_____

_____

5. First workable Internet is create by the US Department of Defence with the name of Advanced Research Projects Agency Network simply ARPANET.

_____

_____

6. What transformed ARPANET in a real global network is the TCP/IP communication model and protocols that are created by Robert Kahn and Vinton Cerf. TCP/IP is created end of the 1970s and integrated into the ARPANET in 1983. This turned ARPANET a real worldwide network called Internet.

_____

_____

_____

_____

_____

_____

Passage 2

Internet Plus: aLife-changing Initiative

Since Premier Li Keqiang **took office**, at least 15 topics regarding Internet Plus have been discussed at State Council executive meetings.

Premier Li **stressed** on many occasions that China should promote the new drivers of **growth** fostered by Internet Plus to create huge potential for a new industrial revolution. ①

New growth drivers fostered by Internet Plus

Internet Plus has covered a wide range of fields-including government services, advanced manufacturing, agriculture and healthcare-and played a role in people's lives in a number of ways.

It provides both individuals and businesses with easy access to government services. It speeds up the building of an industrial Internet. It promotes agricultural improvements, increases income for farmers through the development of rural e-commerce. It has also popularized medical service models such as telemedicine, online appointment and day surgery, and facilitated the sharing of medical information among healthcare institutions. ②

In addition, online learning and education platforms are enabled by big data. Promoting Internet Plus Logistics can develop the new economy while upgrading the traditional economy.

By integrating the Internet with **transportation** as well as Food and Drug Administration, more convenient transportation services are offered and online smart **supervision** is achieved. ③

Measures to support Internet Plus

A slew of measures have been taken to **cultivate** new drivers of growth from Internet Plus. For instance, the government has introduced policies to facilitate faster and more **affordable** Internet connections, canceled policies **hindering** the development of Internet Plus, and built open and sharing platforms. ④

It has also **implemented** the Broadband China Strategy to ensure more than 98 percent of administrative villages have access to broadband Internet by 2020, as well as the New Hardware Project to develop next-generation information technology **infrastructure**. ⑤

Given the features of Internet Plus, efforts have been made to increase government purchases of cloud computing services, encourage innovation in credit products and services, launch **equity** crowdfunding pilot programs, and support Internet startups in going public.

In 2016, China's digital economy reached 22. 58 trillion yuan (about $3. 4 trillion), ranking second globally and accounting for 30. 3 percent of the national gross domestic product. The new economy represented by the digital economy was **thriving**.

The market value of cloud computing and core big data industries registered a **year-on-year** increase of over 30 percent and 45 percent to approximately 50 billion yuan and 16. 8 billion yuan, respectively.

 New Words and Expressions

take office　　上任；就职

stress［stres］　　*v.* 强调；重读

foster［'fɒstə(r)］　　*v.* 促进；培养

transportation［ˌtrænspɔː'teɪʃn］　　*n.* 运输

supervision［ˌsjuːpə'vɪʒn］　　*n.* 监督；管理

cultivate［'kʌltɪveɪt］　　*v.* 栽培；培养

affordable［ə'fɔːdəbl］　　*adj.* 支付得起的；不太昂贵的

hinder［'hɪndə(r)］　　*v.* 阻碍；打扰

implement［'ɪmplɪmənt］　　*v.* 实施；执行

infrastructure［'ɪnfrəstrʌktʃə(r)］　　*n.* 基础；基础设施

equity［'ekwəti］　　*n.* 公平；公正

thrive［θraɪv］　　*v.* 兴旺；繁荣；茁壮成长

year-on-year［经］　　与上年同期数字相比的

Notes

1. Premier Li stressed on many occasions that China should promote the new drivers of growth fostered by Internet Plus to create huge potential for a new industrial revolution.

    Analysis："promote" 意为 "促进"，(potential) 意为 "潜力"。

    Translation：李总理在多个场合强调，中国应推动互联网＋催生的新增长动力，为新的工业革命创造巨大潜力。

2. It has also popularized medical service models such as telemedicine, online appointment and day surgery, and facilitated the sharing of medical information among healthcare institutions.

    Analysis："popularize" 是指 "推广，普及"，"telemedicine" 是指 "远程医疗"，"day surgery" 是指 "日间手术"。

    Translation：互联网＋推广远程医疗、在线预约、日间手术等医疗服务模式，促进医疗机构间医疗信息共享。

3. By integrating the Internet with transportation as well as Food and Drug Administration, more convenient transportation services are offered and online smart supervision is achieved.

    Analysis："integrate" 意为 "融合，整合"，"Food and Drug Administration" 意为 "食品药品监督管理"，"online smart supervision" 意为（在线智能监管）。

    Translation：通过互联网与交通以及食品药品监督管理的融合，提供更便捷的交通服

务，实现在线智能监管。

4. For instance, the government has introduced policies to facilitate faster and more affordable Internet connections, canceled policies hindering the development of Internet Plus, and built open and sharing platforms.

Analysis：“facilitate”意为“促进；帮助”。“platform”意为“平台”。

Translation：例如，政府出台了促进更快、更实惠的互联网连接的政策，取消了阻碍“互联网＋”发展的政策，并建立了开放和共享平台。

5. It has also implemented the Broadband China Strategy to ensure more than 98 percent of administrative villages have access to broadband Internet by 2020, as well as the New Hardware Project to develop next-generation information technology infrastructure.

Analysis：“Broadband China Strategy”意为“宽带中国战略”，“administrative villages”意为“行政村”，“hardware”意为“硬件”。

Translation：它还实施了宽带中国战略，以确保到2020年超过98%的行政村能够接入宽带互联网，以及发展下一代信息技术基础设施的新硬件项目。

 Exercises

Answer the following questions according to the text.

1. Which fields of life covered by Internet Plus are mentioned in the text?

_____

_____

2. How does Internet Plus foster new growth drivers in medical industry?

_____

_____

3. How does Internet Plus foster new growth drivers in agriculture?

_____

_____

4. How much does China's digital economy account for in its GDP in 2016?

_____

_____

5. How much did cloud computing and core big data industries increase year on year?

_____

_____

Passage 3

The Internet Penetration in China

Have you noticed that we can hardly live without mobile internet and smartphone apps in our daily lives?

Actually, the mobile internet has been a **burgeoning** trend in the past decade. Digital living is wonderfully integrated with the real world in China. In 2012, for the first time, the number of mobile internet users **surpassed** the number of people using personal computers to go online.

On September 29, China Internet Network Information Center (CNNIC) **released** the 46th Statistical Report on the Development of China's Internet. The number of Internet users in China had reached 940 million as of June, accounting for one-fifth of the world's total, according to the report. The Internet **penetration** rate in China reached 67 percent, about 5 percentage points higher than the global average, the report said.

The Internet has also been reaching seniors and people in rural areas, gradually bridging the digital divide among the technologically challenged. [1] Last year, 57. 6 percent of the rural population had access to the Internet, while 43. 2 percent of people age 60 and older used the Internet. Residents in 98 percent of China's poor villages had access to the Internet through fiber-optic cables, said the report, noting that the **urban-rural** digital gap had been significantly narrowed. Instant messaging through WeChat and other apps has been the biggest **motivation** for people to get online in the past decade.

In 2013, the number of users seeking online news services saw the highest year-on-year rise than other services, reaching 491 million, slightly more than that for online videos. [②] The number of online video users surpassed that of online news in 2018.

The platform economy is developing rapidly, covering daily activities such as travel, eating and accommodations. The number of people using the Internet to order food deliveries, **hail rides** and book travel services has risen in recent years. Last year, the proportion of people who use the Internet for work purposes and medical services rose 35. 7 percent and 38. 7 percent respectively year-on-year. The number of the country's livestreaming users reached 562 million by the end of June, 309 million of whom are engaged with e-commerce livestreaming, according to the report. Internet services such as online education, medical **consultation** and remote-office **facilities** have great development **potential** due to the influence of the COVID-19 epidemic, said the report, adding that the number of online-education users now exceeds 380 million.

In fact, China's comprehensive and reliable online services have been a silver lining in the country's fight against the novel coronavirus. [③] From take-away deliveries to online shopping **coupons** and livestreaming sales broadcasts, innovation has turned out to be one of the major factors that helped Chinese people tide through the crisis. We can even track and **monitor** our health via QR codes and various health apps.

New Words and Expressions

burgeoning [ˈbɜːdʒənɪŋ] *adj.* 迅速成长的，迅速发展的

surpass [səˈpɑːs] *v.* 超越；胜过

release [rɪˈliːs] *v.* 释放；发表

penetration [ˌpenɪˈtreɪʃn] *n.* 侵入；渗透

urban-rural [ˈɜːbənˈruərəl] *adj.* 城乡的

motivation [ˌməʊtɪˈveɪʃn] *n.* 动机；动力；积极性

hail ride 网约车

consultation [ˌkɒnslˈteɪʃn] *n.* 咨询；请教

facility [fəˈsɪləti] *n.* 设备；设施

potential [pəˈtenʃl] *adj.* 潜在的；可能的

coupon [ˈkuːpɒn] *n.* 票券；礼券

monitor [ˈmɒnɪtə(r)] *v.* 监视；监督；监听

 Notes

1. The Internet has also been reaching seniors and people in rural areas, gradually bridging the digital divide among the technologically challenged.

   Analysis："seniors" 是 "老年人"，"bridge" 是动词，意为 "架桥；缩短…差距"。"digital divide" 意为 "数字鸿沟"，又称为 "信息鸿沟"，也可以用 digital gap 来表示。"the technologically challenged" 意为 "受到技术挑战的（人群）"。

   Translation：互联网也已经普及到老年人和农村地区人口，逐渐弥合了被技术挑战的人群之间的数字鸿沟。

2. In 2013, the number of users seeking online news services saw the highest year-on-year rise than other services, reaching 491 million, slightly more than that for online videos.

   Analysis："year-on-year" 指 "与上年同期数字相比的"，"year-on-year rise" 意为 "同期增长"。

   Translation：2013 年，在所有互联网应用中，网络新闻创增长率新高，用户规模达 4.91 亿，稍微高于看网络视频的用户规模。

3. In fact, China's comprehensive and reliable online services have been a silver lining in the country's fight against the novel coronavirus.

   Analysis："silver lining" 是说乌云被阳光照到时，云周围的那一圈闪光，比喻在失望或不幸中的一线希望，一点慰藉。

   Translation：事实上，中国全面可靠的在线服务一直是中国抗击新型冠状病毒的一线希望。

 计算机专业英语 ···

Exercises

Translate the following short passages into Chinese.

1. Actually, the mobile internet has been a burgeoning trend in the past decade. Digital living is wonderfully integrated with the real world in China. In 2012, for the first time, the number of mobile internet users surpassed the number of people using personal computers to go online.

_____

_____

_____

2. The platform economy is developing rapidly, covering daily activities such as travel, eating and accommodations. The number of people using the Internet to order food deliveries, hail rides and book travel services has risen in recent years.

_____

_____

_____

3. From take-away deliveries to online shopping coupons and livestreaming sales broadcasts, innovation has turned out to be one of the major factors that helped Chinese people tide through the crisis. We can even track and monitor our health via QR codes and various health apps.

_____

_____

_____

Section Ⅲ Grammar

Passive Voice（被动语态）

一、语态概述

语态也是动词的一种形式，表示主语与谓语之间的关系。英语有两种语态：主动语态（active voice）和被动语态（passive voice）。

主动语态表示主语是动作的执行者。例如：Many people speak English. 谓语 speak 的动作是由主语 many people 来执行的。

被动语态表示主语是动作的承受者，即行为动作的对象。例如：English is spoken by many people. 主语 English 是动词 speak 的承受者。

二、被动语态的构成

被动语态由"助动词 be + 及物动词的过去分词"构成。人称、数和时态的变化是通过 be 的变化表现出来的。

现以动词 give 为例，其被动语态的各种时态如下：

时＼态	一　般	进　行	完　成
现在	am/is/are + given	am/is/are + being + given	have/has + been + given
过去	was/were + given	was/were + being + given	had + been + given
将来	shall/ will + be + given	shall/will + beinggiven	shall/will + have + been + given
含情态词的	can/may/must + be + given		

三、被动语态的特殊结构形式

1. 带情态动词的被动结构。其形式为：情态动词 + be + 过去分词。

例：The baby should be taken good care of by the baby-sitter.

2. 有些动词可以有两个宾语，在用于被动结构时，可以把主动结构中的一个宾语变为主语，另一宾语仍然保留在谓语后面。通常变为主语的是间接宾语。

例：His mother gave him a present for his birthday. 可改为 He was given a present by his mother for his birthday.

3. 当"动词 + 宾语 + 宾语补足语"结构变为被动语态时，将宾语变为被动结构中的主语，其余不动。

例：Someone caught the boy smoking a cigarette. 可改为 The boy was caught smoking a cigarette.

4. 在使役动词 have，make，get 以及感官动词 see，watch，notice，hear，feel，observe 等后面不定式作宾语补语时，在主动结构中不定式 to 要省略，但变为被动结构时，要加 to。

例：Someone saw a stranger walk into the building. 可改为 A stranger was seen to walk into the building.

5. 有些相当于及物动词的动词词组，如"动词 + 介词"，"动词 + 副词"等，也可以用于被动结构，但要把它们看作一个整体，不能分开。其中的介词或副词也不能省略。

例：The meeting is to be put off till Friday.

6. 非谓语动词的被动语态。

v. + ing 形式及不定式 to do 也有被动语态（一般时态和完成时态）。

例：I don't like being laughed at in the public.

四、It is said that + 从句及其他类似句型

一些表示"据说"或"相信"的动词如 believe, consider, expect, report, say, suppose, think 等可以用于句型"It + be + 过去分词 + that 从句"或"主语 + be + 过去分词 + to do sth."。有：

It is said that... 据说   It is reported that...据报道

It is believed that...大家相信   It is hoped that...大家希望

It is well known that...众所周知   It is thought that...大家认为

It is suggested that...据建议

例：It is said that the boy has passed the national exam. （= The boy is said to have passed the national exam.）

五、谓语动词的主动形式表示被动意义

1. 英语中有很多动词如 break, catch, clean, drive, lock, open, sell, read, write, wash 等，当它们被用作不及物动词来描述主语特征时，常用其主动形式表达被动意义，主语通常是物。

例：This kind of cloth washes well.

注意：主动语态表被动强调的是主语的特征，而被动语态则强调外界作用造成的影响。

试比较：The door won't lock。（指门本身有毛病）

The door won't be locked.（指不会有人来锁门，指"门没有锁"是人的原因）

2. 表示"发生、进行"的不及物动词和短语，如：happen, last, take place, break out, come out, come about, come true, run out, give out, turn out 等以主动形式表示被动意义。

例：How do the newspapers come out? 这些报纸是如何引出来的呢？

3. 系动词没有被动形式，但有些表示感受、感官的连系动词 feel, sound, taste, book, feel 等在主系表结构中常以主动形式表示被动意义。

例：Your reason sounds reasonable.

六、非谓语动词的主动形式表示被动意义

在某些句型中可用动名词和不定式的主动形式表被动意义。

1. 在 need, want, require, bear 等词的后面，动名词用主动形式表示被动意义，其含义相当于动词不定式的被动形式。The house needs repairing (to be repaired). 这房子需要修理。

2. 形容词 worth 后面跟动名词的主动形式表示被动含义，但不能跟动词不定式；而 worthy 后面跟动词不定式的被动形式。

例：The picture-book is well worth reading. （= The picture-book is very worthy to be read. ）

3. 动词不定式在名词后面作定语，不定式和名词之间有动宾关系时，又和句中另一名词或代词构成主谓关系，不定式的主动形式表示被动含义。

例：I have a lot of things to do this afternoon. （to do 与 things 是动宾关系，与 I 是主谓关系。）

试比较：I'll go to the post office. Do you have a letter to be posted? （此处用不定式的被动语态作定语表明 you 不是 post 动作的执行者。）

4. 在某些"形容词＋不定式"做表语或宾语补足语的结构中，句子的主语或宾语又是动词不定式的逻辑宾语时，这时常用不定式的主动形式表达被动意义。这些形容词有 nice, easy, fit, hard, difficult, important, impossible, pleasant, interesting 等。

例：This problem is difficult to work out. （可看作 to work out 省略了 for me）

5. 在 too…to…结构中，不定式前面可加逻辑主语，所以应用主动形式表示被动意义。

例：This book is too expensive （for me） to buy.

6. 在 there be…句型中，当动词不定式修饰名词作定语时，不定式用主动式作定语，重点在人，用被动形式作定语，重点在物。

例：There is no time to lose （to be lost）. （用 to lose 可看成 for us to lose；用 to be lost, lost time 的主体不明确。）

7. 在 be to do 结构中的一些不定式通常应用主动表主动，被动表被动。然而，受古英语的影响，下列动词 rent, blame, let 等仍用不定式的主动形式表示被动意义。

例：Who is to blame for starting the fire?

Exercises

Choose the best answer.

1. The house, which _____ last night, _____ my aunt but she doesn't live there any more.

    A. was broken into; is belonged to   B. broke into; is belonged to

    C. broke into; belonging to   D. was broken into; belongs to

2. The singer's music video _____ nearly 9 million times since it was posted on the Internet four weeks ago.

    A. viewed   B. has viewed

    C. was viewed   D. has been viewed

3. Now the world's attention _____ the stocking markets, as they have great influence on the world's economy.

    A. is fixing on   B. is being fixed on

    C. has fixed on   D. had been fixed on

4. Local governments _____ to strengthen water transport safety management _____ recent fatal accidents.

A. are urging, followed   B. are being urged, following

C. are urged, to follow   D. urge, being followed

5. A human case of H7N9 was reported in 2014 when a woman _____ to be infected with the bird flu virus.

    A. confirmed   B. had been confirmed

    C. was confirmed   D. have confirmed

6. In no time _____ by a string of measures backing Hainan's efforts to deepen reform and opening-up.

    A. the landmark decision was followed

    B. was the landmark decision followed

    C. did the landmark decision follow

    D. the landmark decision had been followed

7. Nowadays, cycling, along with jogging and swimming, _____ as one of the best all-round forms of exercise.

    A. regard   B. is regarded   C. are regarded   D. regards

8. They are trying to make sure that 5G terminals _____ by 2022 for the Beijing Winter Olympics.

    A. will install   B. will have been installed

    C. are installed   D. have been installed

9. This company _____ roundly by unionists and social justice groups when it fired a number of workers for no reason a decade ago.

    A. condemned   B. has condemned

    C. has been condemned   D. was condemned

10. Time magazine has chosen "The Guardians", a group of journalists who _____ for their work, as Person of the Year, for taking great risks in pursuit of greater truths.

    A. will target   B. have targeted

    C. will be targeted   D. have been targeted

Section Ⅳ Writing

Letter of Complaint（投诉信）

投诉信是在日常生活中或商务活动中对所遇到的困扰、恶劣的服务态度、劣质的产品等向对方或有关单位提出投诉、抗议和改进的建议时书写的信。投诉信是对服务或产品的质量表示不满，写信时要把不满意的理由说清楚，语气要肯定，但是不宜过于生硬或使用伤害对方的语言。

Part 1 Sample

The following is a letter of complaint. Please read and try to understand it.

Dear General Manager,

    I'm writing because I need to complain to you about what happened to me at your store last week.

    I was in Boluo last week for a conference. The evening before I left for home, I went to your store and bought a shirt for my son. The shop assistant helped me choose one and I trusted her judgement and left without check. After I got back to the hotel, I found a small hole on the collar. So I came back to exchange it immediately, but she wouldn't exchange it for me and said she could do nothing about it. I know that your store is one of the best stores in the country. But why couldn't a simple customer's service like that be done?

    I hope that the store can answer me and exchange the shirt.

    I am looking forward to your reply.

Sincerely,

Jack

Part 2 Template

从上面的样例可以看出英语投诉信（Letter of Complaint）的书写规则与其他书信的书写规则基本相同。在写投诉信时切忌一味恼怒，一定要做到有理有据，写这类信时要注意以下几点：

首段：表明来信所要投诉的问题，尽可能做到客观礼貌，给读者留下好印象，应该记住读信人不一定就是错误的责任人，他的合作对事情的最后解决有着非常重要的作用。

主体段落：写明投诉的原因、问题的经过及产生的后果。可以说具体的理由，也可以说问题的具体体现方式。

结尾段：提出解决方案，不需展开。这段应体现书信的礼貌原则，可以用类似"如果我能…，我将十分感激"这样的表达方式。还要注意做到公平公正。

投诉信应重点表明投诉的原因，叙事应客观、准确、简洁。最后提出的解决方案应切实可行。在表达自己的不满时，语言要把握分寸，不失风度。

Dear _____,

    I am _____ . I am sorry to trouble you but I am afraid that I have to make a complaint about _____ .（首段进行自我介绍，并说明要投诉的地方）

    The reason for my dissatisfaction is _____ . In the first place, _____ . In addition, _____ . Under these circumstances, I find it _____ .（第二段采用总分总的结构，先总体表达不满的原因，再分条具体说明一下投诉原因，最后总结一下以上几点给自己带来的不便）

    I appreciate it very much if you could _____ , and I would like to have this matter settled by _____ . Thank you for your consideration and I will be looking forward to your reply.（结尾段提出自己的建议和诉求，设定解决问题的期限并表达感谢）

Sincerely yours,
Li Hua

Part 3　Useful Patterns

1. I'm writing because I need to complain to you about...

2. I'm sorry to point out the defect in the air-conditioner.

3. We wish to make the following complaint to you and hope...will do something about it.

4. I was shocked to find...（product）purchased on...（date）at...（place）by us did not function well.

5. I would appreciate it very much if you could give me the refund.

6. I strongly insist that you change...as soon as possible.

7. I would like to draw attention to...

8. I should be very obliged if you would look into this matter as soon as possible.

9. I'm looking forward to an early reply.

10. We will appreciate your willingness to make up for the loss.

1. 我写这封信是想向你投诉…

2. 对不起，我得指出空调的毛病。

3. 我们向你抱怨以下内容，希望你能给予解决。

4. 我惊讶地发现，我们…日在…购买的…工作不好。

5. 如能退款，不胜感激。

6. 我强烈要求尽快换…

7. 我希望你们注意…

8. 如果您能尽快调查此事，我将不胜感激。

9. 我期待早日回复。

10. 我们真的希望你们能对损失作出补偿。

Part 4   Exercises

Ⅰ. Translate the following letter of complaint into Chinese.

Dear manager,

   I write this letter to you to make some complaints about the computer I bought in your store yesterday afternoon. There's something wrong with it. That makes me extremely unhappy.

   The computer cannot be properly shut down when I got it back to the office. When I click the shutdown button, it seems that the machine gives no response. And I'm so annoyed with it.

   It's obvious that you didn't carefully examine the machine before you sold it. I think your store should take full responsibility for selling me the defective machine. I insist that you give me a satisfactory reply. I do want you to give back my money as soon as possible.

Sincerely yours,

Jack

Ⅱ. Write a letter of complaint according to the information given in Chinese.

假设你是李华，你对昨天乘坐的某次航班非常不满，理由如下：

1. 航空公司弄丢了你的一个包，包内有秘密文件；

2. 飞机晚点一个半小时；

3. 飞机上的食物很差；

4. 空姐的服务态度不好。

请就此事给航空公司的经理写一封投诉信。

注意：1. 词数 100 左右。

2. 信的开头和结尾已给出，不计入总字数。

Unit 6

Big Data

Unit Goals

After learning this unit, you will be able to:

- talk about big data
- understand the passages and grasp the key words and expressions
- review the grammar of infinitive
- write a letter of apology

Section Ⅰ   Listening & Speaking

Part A   Words and Expressions

personal information 个人信息	personnel transfer 人员流动
privacy security 隐私安全	vehicle movement 车辆流动
family information 家庭信息	logistics information 物流信息
ID number 身份证号码	product recommendations 产品推荐
transparent 透明的；易识破的	payment model 支付方式
go into effect 生效	peak hour 高峰时段
health code 健康码	online duration 在线时长
travel card 行程卡	log in 登录
nucleic acid test 核酸检测	user need 用户需求
high-risk area 高风险区域	big data analyst 大数据分析师
novel coronavirus 新冠病毒	

Part B   Conversations

Read the conversations carefully and then complete the communicative tasks.

Conversation 1

A：Xiao Li, have you seen a recent **sitcom** *The Lord of Losers*?

B：No, is it interesting? About what?

A：It's funny. It mainly talks about several young people who have experienced a series of **ironic** things in the workplace.

B：Oh, then the play should be very close to our life.

A：Yes. In yesterday's story the "Trivial Department" was deeply involved in the routine of advertising due to the privacy leakage, and was **harassed** by various pushes every day.

B：Indeed, such problems are even more **frequent** and **outrageous** in life.

A: For example, one minute ago, you just mentioned something when you were chatting with someone, and the next minute, when you open the online shopping software, and it happens to push that item; another example is that you browse something on the website, and then they send you related short videos...

B: There is also the situation that users' personal information is stolen. Swindlers can even tell your name, ID number, hobbies, family information, recent status, etc.

A: Unknowingly, personal information of the public may be stolen and used, and they fall into the traps of Telecom fraud and online loan, which is hard to prevent.

B: Technology brings infinite convenience to people, but problems follow: under big data, privacy is becoming more transparent, and the information left by individuals in major APPs or devices is stolen, and even secretly sold.

A: How can we protect the privacy of the public?

B: We should take measures in legislation, management, technology and other aspects to protect personal privacy security.

A: Is there a law on privacy security in our country?

B: Yes. The Data Security Law went into effect on September 1, 2021, followed by the Personal Information Protection Law two months later.

A: Great. Only by increasing the cost of crime and strengthening the law and management can we solve the problem fundamentally.

B: There is still a long way to go to resolve the conflict between big data and privacy.

 Key Words

sitcom	情景喜剧
ironic	令人啼笑皆非的
harass	使烦扰，折磨
frequent	频繁的
outrageous	无法容忍的；反常的

Task 1 Act out the conversation with your partner based on the following clues.

A: Xiao Li, have you seen a recent sitcom *The Lord of Losers*?

B: No, is it interesting? About what?

A: It's funny. It mainly talks about several young people who have experienced a series of ironic things in the workplace.

B: Oh, then the play should be very close to our life.

A: Yes. In yesterday's story the "Trivial Department" was deeply involved in the routine of advertising due to the privacy leakage, and _____①_____.

B: Indeed, _____②_____.

A: For example, one minute ago, you just mentioned something when you were chatting with someone, and the next minute, when you open the online shopping software, and it happens to push that item; another example is that you browse something on the website, and then they send you related short videos...

B: There is also the situation that users' personal information is stolen. Swindlers can even tell your name, ID number, hobbies, family information, recent status, etc.

A: Unknowingly, personal information of the public may be stolen and used, and ____③____, which is hard to prevent.

B: Technology brings infinite convenience to people, but problems follow: under big data, privacy is becoming more transparent, and the information left by individuals in major APPs or devices is stolen, and even secretly sold.

A: How can we protect the privacy of the public?

B: _____④_____.

A: Is there a law on privacy security in our country?

B: Yes. The Data Security Law went into effect on September 1 last year, followed by the Personal Information Protection Law two months later.

A: Great. _____⑤_____.

B: There is still a long way to go to resolve the conflict between big data and privacy.

Conversation 2

A: Hi, Xiao Li. How was your holiday?

B: Just so so. I've been indoors most of the time because of the coronavirus.

A: Why are you staying home? You can even travel with a health code of **negative** nucleic acid test and a travel card as long as you avoid high-risk areas.

B: Is the health code so amazing that it can help us reduce the risk of being **infected** by novel coronavirus?

A: Of course, this is the age of big data!

B: What is big data?

A: Big data refers to data sets that exceed the collection, storage, management and analysis capabilities of traditional database tools.

B: Is big data widely used in our daily lives?

A: Big data has been **integrated** into our lives. For example, with the rapid development of the Internet, and the **advent** of smartphones and "wearable" computing devices, every change in our behavior, our location, and even our physical data becomes the data that can be recorded and analyzed.

B: Big data technology must have played an important role in the prevention and control of the

epidemic.

A： Yes. With the help of big data, the government can query information at any time, such as personnel transfer, vehicle movement, logistics information and other relevant information; the public can **keep abreast of** the relevant information, the latest trends and development of the epidemic.

B： What are the other applications of big data?

A： The simplest example is that every day when we open a shopping website, some product recommendations will pop up on both sides of the page. These recommendations are precisely because big data provides the merchants and advertisers with the **commodity** information you have searched on Taobao, Tmall and other trading websites, so as to push the products related to you.

B： No wonder shopping websites always recommend things that interest me.

A： The most essential application of big data is prediction, that is, to analyze certain characteristics from massive data, and then predict what may happen in the future. With the data "big" enough, almost all the needs in your life can be predicted.

B： Big data does bring a lot of convenience to our lives.

 Key Words

negative　阴性的；消极的；否定的

infect　传染；使感染

integrate　合并；成为一体；融入

advent　出现，到来，问世

epidemic　流行病

keep abreast of　（使）保持与…并列，了解…的最新情况

commodity　商品

Task 2　Act out the conversation with your partner based on the following clues.

A： Hi, Xiao Li. How was your holiday?

B： Just so so. I've been indoors most of the time because of the coronavirus.

A： Why are you staying home? You can even travel with a health code of negative nucleic acid test and a travel card as long as avoid high-risk areas.

B： Is the health code so amazing that it can help us reduce the risk of being infected by novel coronavirus?

A： Of course, this is the age of big data!

B： What is big data?

A： ＿＿＿＿①＿＿＿＿.

B: Is big data widely used in our daily lives?

A: _____②_____ . For example, with the rapid development of the Internet, and the advent of smartphones and "wearable" computing devices, every change in our behavior, our location, and even our physical data becomes the data that can be recorded and analyzed.

B: _____③_____ .

A: Yes. With the help of big data, the government can query information at any time, such as personnel transfer, vehicle movement, logistics information and other relevant information; _____④_____ .

B: What are the other applications of big data?

A: The simplest example is that every day when we open a shopping website, some product recommendations will pop up on both sides of the page. These recommendations are precisely because big data provides the merchants and advertisers with the commodity information you have searched on Taobao, Tmall and other trading websites, so as to push the products related to you.

B: No wonder shopping websites always recommend things that interest me.

A: _____⑤_____ . With the data "big" enough, almost all the needs in your life can be predicted.

B: Big data does bring a lot of convenience to our lives.

Conversation 3

A: Ada, what are you doing? That's the biggest **spreadsheet** I've ever seen!

B: Hi, Weiwei. Yeah, just you wait. I'm developing a product that will **revolutionize** English teaching!

A: Great! So why are you looking at screens of numbers?

B: Not numbers, data. Big data! This is what I need.

A: How do you plan to use this data?

B: I haven't figured it out yet. But that's not important. Everyone's talking about big data.

A: Big data is indeed a hot topic. But big data needs to be collected and **analyzed**, and you have to know how to handle this data. So, Ada. How's your research going now?

B: Umm... not well. It's hard to know where to start. Do you have any suggestions?

A: I think you might need some information such as user's payment model for language learning applications, peak hours for user login, use's online **duration** after logging in, changes in user needs at different language levels. . .

B: Stop, stop, stop. I'm trying to write them down. Weiwei?

A: Yes?

B: Can I hire you as a big data analyst?

A: Sorry, Ada, you're too late. I'm launching a product of my own. It helps **predict** football scores.

B: Ah! That sounds great. I wish you success.

Key Words

spreadsheet 电子表格

revolutionize 使彻底变革

analyze 分析；分解；化验

duration 持续时间；期间

analyst 分析者；化验员

predict 预测

Task 3 Act out the conversation with your partner based on the following clues.

A: Ada, what are you doing? That's the biggest spreadsheet I've ever seen!

B: Hi, Weiwei. Yeah, just you wait. I'm developing a product that will revolutionize English teaching!

A: Great! So why are you looking at screens of numbers?

B: Not numbers, data. Big data! This is what I need.

A: How do you plan to use this data?

B: _____①_____ . But that's not important. Everyone's talking about big data.

A: Big data is indeed a hot topic. _____②_____ . So, Ada. How's your research going now?

B: Umm...not well. It's hard to know where to start. _____③_____ ?

A: I think you might need some information such as user's payment model for language learning applications, peak hours for user login, user's online duration after logging in, changes in user needs at different language levels. . .

B: Stop, stop, stop. I'm trying to write them down. Weiwei?

A: Yes?

B: _____④_____ ?

A: Sorry, Ada, you're too late. _____⑤_____ . It helps predict football scores.

B: Ah! That sounds great. I wish you success.

Part C Passages

Listen to the following passages carefully and fill in the blanks with the information you've heard.

听力材料

Passage 1

Big data refers to _____ whose size _____ the ability of conventional _____ tools to acquire, store, manage, and analyze. The current big data refers to a huge data _____, which refers to a collection of data that cannot be captured, managed and processed with conventional _____ tools within a certain time frame.

Passage 2

"The app and database will help us _____ more precise and well-rounded data on the population, _____, ages, gender ratio, birth and deaths of _____, who live in deep mountains and are hard to _____," said Chen Peng, a researcher with the base who _____ a paper on "Giant Panda Face Recognition Using Small Database".

Passage 3

Cherry blossoms _____ the fragility of life in Japanese _____ as full blooms only last about a week before the petals start falling off trees. And in that period and the _____ weeks, shops will _____ their shelves with sakura-themed merchandise. Pink and white blossoms seem to _____ everything from beer cans to sakura-flavored chips and flower-themed candy.

Passage 4

The application of big data is also very _____. All walks of life in society, including finance, automobiles, catering, _____, energy, physical fitness, and entertainment, have been _____ the footprint of big data. Like the financial industry, big data plays a major role in the three major financial _____ fields of high-frequency trading, social sentiment analysis, and credit risk analysis. In the automotive industry, _____ cars using big data and Internet of Things technology will enter our daily lives in the not-too-distant future. The Internet industry, with the help of big data technology, can analyze customer behavior, carry out product recommendation and targeted advertising and other applications. In general, big data has become the most important part of the information age.

Passage 5

The most basic element of _____ when the delicate pink and white petals will begin to unfurl is a large _____ of temperatures. That's because the flowers will come earlier if temperatures rise quickly in spring. _____, if temperatures in the autumn and winter period are higher than usual, the blooms can end up being delayed. _____ weather can affect the trees too, with unusual patterns in 2018 prom _____ some blossoms to appear in October, well before the usual season. In general, blooms begin as early as March in southern Kyushu and appear as late as May in northernmost Hokkaido.

Passage 1

What is Big Data?

For people living in the information age, we believe that the term "big data" is already familiar to us. We have been **injected** into our computer network by various kinds of data. With the explosive growth of data, big data has **emerged** as the times require.① Then what is big data in the end?

The concept of big data was formally **proposed** in the "Nature" magazine published in the United States in September 2008, and then on February 1, 2011, the American "Science" magazine analyzed the **impact** of big data on peolpe's lives for the first time through social surveys. May 2011, McKinsey **Institute** distributed the report. Big data refers to data sets whose size exceeds the ability of conventional database tools to acquire, store, manage, and analyze.② The current big data refers to a huge data collection, which refers to a collection of data that cannot be captured, managed and processed with **conventional** software tools within a certain time frame.③ It requires a new processing model to have stronger decision-making power, **insight** and discovery. And the massive, high growth rate and **diversified** information assets of process **optimization** capabilities. ④

　　The application of big data is also very extensive. All walks of life in society, including finance, automobiles, catering, telecommunications, energy, physical fitness, and entertainment, have been integrated into the footprint of big data. ⑤ Like the financial industry, big data plays a major role in the three major financial innovation fields of high-frequency trading, social sentiment analysis, and credit risk analysis. In the automotive industry, driverless cars using big data and Internet of Things technology will enter our daily lives in the not-too-distant future. The Internet industry, with the help of big data technology, can analyze customer behavior, carry out product recommendation and targeted advertising and other applications. In general, big data has become the most important part of the information age.

 New Words and Expressions

　　inject [ɪnˈdʒekt]　　v. 注射;(给…)添加,增加(某品质)

　　emerge [ɪˈmɜːdʒ]　　v. 浮现,出现;显露

　　propose [prəˈpəʊz]　　v. 提出(理论或解释);提议,建议

　　impact [ˈɪmpækt]　　n. 巨大影响;撞击,冲击力

　　survey [ˈsɜːveɪ]　　n. 考察,调查

　　institute [ˈɪnstɪtjuːt]　　n. 研究所,学院,协会

　　conventional [kənˈvenʃən(ə)l]　　adj. 依照惯例的,遵循习俗的

　　insight [ˈɪnsaɪt]　　n. 洞悉,了解;洞察力

　　diversify [daɪˈvɜːsɪfaɪ]　　v. (使)多样化,(使)不同

　　optimization [ˌɒptɪmaɪˈzeɪʃən]　　n. 最佳化,最优化

 Notes

1. We have been injected into our computer network by various kinds of data. With the explosive growth of data, big data has emerged as the times require.

　　Translation:各种各样的数据被注入我们的计算机网络中。随着数据的爆炸式增长,

大数据应运而生。

2. Big data refers to data sets whose size exceeds the ability of conventional database tools to acquire, store, manage, and analyze.

Translation：大数据是指其大小超出了常规数据库工具获取、存储、管理和分析能力的数据集。

3. The current big data refers to a huge data collection, which refers to a collection of data that cannot be captured, managed and processed with conventional software tools within a certain time frame.

Analysis："data collection" 意为 "数据集合"，"software tools" 意为 "软件工具"，"time frame" 意为 "时间范围"。

Translation：而现在的大数据指的是巨量数据集合，指无法在一定时间范围内用常规软件工具进行捕捉、管理和处理的数据集合。

4. It requires a new processing model to have stronger decision-making power, insight and discovery. And the massive, high growth rate and diversified information assets of process optimization capabilities.

Analysis：It 指代前文提到的 "the current big data" 现在的大数据。"decision-making power" 意为 "决策力"，"insight" 意为 "洞察力"，"massive" 意为 "庞大的，海量的"，"high growth rate" 意为 "高增长率"，"assets" 意为 "资产，有价值的人或物"。

Translation：它需要一种新的处理模式，以便具有更强的决策能力、洞察力和发现力。以及海量、高增长率及具有流程优化能力的多样化信息资产。

5. All walks of life in society, including finance, automobiles, catering, telecommunications, energy, physical fitness, and entertainment, have been integrated into the footprint of big data.

Analysis："All walks of life" 意为 "各行各业，各界人士"，"integrate into" 意为 "成为一体，融入"。

Translation：社会各行各业，包括金融、汽车、餐饮、电信、能源、健身和娱乐，都融入了大数据的足迹。

Exercises

Ⅰ. Answer the following questions according to the text.

1. When and where was the concept of big data first proposed formally?

_____

_____

2. What is big data?

_____

3. Is big data widely used in our life?

_____

_____

4. What kinds of financial fields in which big data is applied are mentioned in the text?

_____

_____

5. What is the application of big data in the automotive industry?

_____

_____

Ⅱ. Fill in the blanks with words according to the meaning of the article by memory.

The application of big data is also very _____. _____ in society, including finance, automobiles, catering, telecommunications, energy, physical fitness, and entertainment, have been _____ the footprint of big data. Like the financial industry, big data plays a major role in the three major financial _____ fields of high-frequency trading, social sentiment analysis, and credit risk analysis. In the automotive industry, _____ using big data and Internet of Things technology will enter our daily lives in the _____ future. The Internet industry, with the help of big data technology, can _____ customer behavior, _____ product recommendation and targeted advertising and other applications. In general, big data has become the most important part of the information age.

Ⅲ. Fill in the blanks with the words given below. Change the forms when necessary.

emerge	propose	survey
exceed	convention	strong
apply	drive	

1. In the automotive industry, _____ cars using big data and Internet of Things technology will enter our daily lives in the not-too-distant future.

2. It requires a new processing model to have _____ decision-making power, insight and discovery.

3. Big data refers to data sets whose size _____ the ability of conventional database tools to acquire, store, manage, and analyze.

4. The concept of big data was formally _____ in the "Nature" magazine published in the United States in September 2008.

5. With the explosive growth of data, big data has _____ as the times require.

6. On February 1, 2011, the American "Science" magazine analyzed big data for the first time through social _____ .

7. The current big data refers to a huge data collection, which refers to a collection of data that cannot be captured, managed and processed with _____ software tools within a certain time frame.

8. The _____ of big data is also very extensive.

Ⅳ. Translate the following sentences into Chinese.

1. For people living in the information age, we believe that the term "big data" is already familiar to us.

_____

_____

2. Big data refers to data sets whose size exceeds the ability of conventional database tools to acquire, store, manage, and analyze.

_____

_____

3. The current big data refers to a huge data collection, which refers to a collection of data that cannot be captured, managed and processed with conventional software tools within a certain time frame.

_____

_____

4. All walks of life in society, including finance, automobiles, catering, telecommunications, energy, physical fitness, and entertainment, have been integrated into the footprint of big data.

_____

_____

5. The Internet industry, with the help of big data technology, can analyze customer behavior, carry out product recommendation and targeted advertising and other applications.

_____

_____

Passage 2

Giant Panda Facial Recognition App Has Introduced

Having trouble **discerning** one furry panda from another? A facial recognition app will make it easy for you. The app is developed by the Chengdu Research Base of Giant Panda Breeding along with researchers in Singapore Nanyang Technological University and Sichuan Normal University. ①
Visitors to the panda base in Chengdu the **capital** of Sichuan, one of China's top tourist attractions, will be able to scan the panda's face with the mobile app to get information about that particular bear. ②

The image analysis research **kicked off** in 2017. A database now contains about 120,000 images and 10,000 video clips of giant pandas. Close to 10,000 panda pictures have been analyzed, marked and **annotated**. Using the database, researchers are able to carry out automatic facial recognition on panda faces to tell one animal from another, the center said. It's not just a **gimmick** for tourists, though. Researchers say the technology will help them analyze data on pandas both in **captivity** and the wild.

由于大熊猫的照片通常是在不受约束的情况下拍摄的，所以质量、清晰度等参差不齐，大熊猫"认脸"的算法也就比较复杂。大致来说包括几大关键元素：姿势、表情、眼睛、耳朵、面部阴影、嘴部咬合等

姿势　表情　面部阴影
耳朵　眼睛　嘴部咬合

"The app and database will help us gather more precise and **well-rounded** data on the population, distribution, ages, gender ratio, birth and deaths of wild pandas, who live in deep mountains and are hard to **track**," said Chen Peng, a researcher of the base who co-authored a paper on "Giant Panda Face Recognition Using Small Database."③ "It will definitely help us improve efficiency and effectiveness in **conservation** and management of the animals," Chen said.

China has carried out four scientific **field research** projects of giant pandas in the wild. The giant panda was scientifically discovered 150 years ago and named in the city of Ya'an, Sichuan. It remains one of the world's most **endangered** species. The number of captive pandas was 548 globally as of November last year. Fewer than 2,000 pandas live in the wild, mostly in the provinces of Sichuan and Shaanxi.

New Words and Expressions

discern [dɪˈsɜːn]　*v.* (艰难地或努力地)看出，觉察出

capital [ˈkæpɪt(ə)l]　*n.* 省会；首都

kick off　开始

annotate [ˈænəteɪt]　*v.* 注解，注释

gimmick [ˈgɪmɪk]　*n.* 骗人的玩意；花招

captivity［kæpˈtɪvəti］　*n.* 圈养；被俘；囚禁

well-rounded［ˌwelˈraʊndɪd］　*adj.* 全面的；面面俱到的；丰满的

track［træk］　*v.* 追踪

conservation［ˌkɒnsəˈveɪʃ(ə)n］　*n.* 保护；保存

field research　实地调查研究

eNDANgered［ɪnˈdeɪndʒəd］　*adj.* (动植物)濒危的

Notes

1. The app is developed by the Chengdu Research Base of Giant Panda Breeding along with researchers in Singapore Nanyang Technological University and Sichuan Normal University.

   Analysis："along with" 意为 "与…一道"，"normal" 意为 "常态，通常标准"，"Normal University" 意为 "师范大学"。

   Translation：这款应用由成都大熊猫繁育研究基地、新加坡南洋理工大学和四川师范大学的研究人员共同开发。

2. Visitors to the panda base in Chengdu the capital of Sichuan, one of China's top tourist attractions, will be able to scan the panda's face with the mobile app to get information about that particular bear.

   Translation：前往熊猫基地参观的游客只要使用该应用程序扫描熊猫面部，就能获知每只熊猫的信息。该基地位于四川省会成都，中国最著名的旅游胜地之一。

3. "The app and database will help us gather more precise and well-rounded data on the population, distribution, ages, gender ratio, birth and deaths of wild pandas, who live in deep mountains and are hard to track," said Chen Peng, a researcher with the base who co-authored a paper on "*Giant Panda Face Recognition Using Small Database*".

   Analysis："co-author" 是指 "合著者，共同创作者"。co-前缀表示 "一起；共同；联合"，如 co-produce 共同制作，co-chairman 联合主席，co-founder 共同创始人。

   Translation：论文《基于小数据集的大熊猫个体识别》作者之一、成都大熊猫繁育研究基地研究员陈鹏说："这款应用和数据库将帮助我们收集有关野生大熊猫数量、分布、年龄、性别比例、出生和死亡的更准确全面的数据。大熊猫们生活在深山中，很难追踪。"

Exercises

Answer the following questions according to the text.

1. Who developed the Panda Facial Recognition App?

   _____

   _____

   _____

   _____

2. How many panda images and video clips are there in the database?

_____

_____

3. How can the visitors to the panda base tell one panda from another?

_____

_____

4. How many captive giant pandas are there around the world as of November last year?

_____

_____

5. Where are China's wild giant pandas mainly distributed?

_____

_____

Passage 3

Prediction of Cherry Blossom

As spring approaches in Japan, the country's weather forecasters face one of their biggest missions of the year: predicting exactly when the cherry blossoms will bloom. Japan's Sakura or cherry blossom season is **feverishly anticipated** by locals and visitors alike. Many tourists plan their entire trips around the blooms, and Japanese flock to parks in their millions to enjoy the seasonal **spectacle**. "People pay more attention to the cherry blossom season than any other flower in Japan," Ryo Dojo, an official of the statistics unit at the Japan Meteorological Agency, told AFP.

The most basic element of predicting when the **delicate** pink and white petals will begin to **unfurl** is a large data set of temperature. That's because the flowers will come earlier if the temperature rises quickly in spring. Conversely, if the temperature in the autumn and winter period is higher than usual, the blooms can end up being delayed. Extreme weather can affect the trees too, with unusual patterns in 2018 prompting some blossoms to appear in October, well before the usual season. In general, blooms begin as early as March in southern Kyushu and appear as late as May in northernmost Hokkaido.

In a bid to improve its forecasts, some outfits have started crowdsourcing data, including Weathernews, a firm in Chiba near Tokyo. [1] It relies on photos of buds sent in regularly by 10,000 citizens across the country who are **registered** on the company's website and app. "Cherry blossom forecasting is impossible for us without this system," a spokeswoman said.

The company launched what they call the "Sakura Project" in 2004, signing up members who choose their own cherry tree and send pictures of its buds to the firm at regular intervals. [2] Just observing the bud can give surprisingly accurate information about how far the flower is from full bloom. A sakura bud still a month from blossoming will be small and firm, but after 10 days, the tip turns slightly yellow-green, and then a darker green part emerges. When the tip of the bud turns a faint pink, it's just a week until bloom-time. Thanks to the project, Weathernews has **accumulated** data from two million reports on cherry flower buds in the past 15 years, which it uses to increase the accuracy of its forecasting.

It also **incorporates** weather data collected from its own observation devices across Japan-13,000 locations in total, 10 times more than the official weather agency has. [3] Weathernews employees also call around 700 parks regularly to check the growth of cherry flower buds. The company and other forecasters also employ **mathematical models** and **algorithms**. Otenki Japan, a forecaster run by a **subsidiary** of precision-equipment manufacturer Shimadzu, even began using **artificial intelligence** to predict cherry blossoms in 2018.

The forecasts are not only for flower fans but reflect the fact that Sakura season is big business in Japan. Cherry blossoms symbolize the fragility of life in Japanese culture as full blooms only last about a week before the petals start falling off trees. And in that period and the preceding weeks, shops will pack their shelves with sakura-themed merchandise. Pink and white blossoms seem to decorate everything from beer cans to sakura-flavored chips and flower-themed candy.

New Words and Expressions

feverishly [ˈfiːvərɪʃli] *adv.* 狂热地，兴奋地

anticipate [ænˈtɪsɪpeɪt] *v.* 期望；预期，预料

spectacle [ˈspektək(ə)l] *n.* 壮观的景象；精彩的表演

delicate [ˈdelɪkət] *adj.* 纤弱的；脆弱的；精致的

unfurl [ʌnˈfɜːl] *v.* 展开；使…临风招展

register [ˈredʒɪstə(r)] *v.* 登记，注册

accumulate [əˈkjuːmjəleɪt] *v.* 积累，积攒

incorporate [ɪnˈkɔːpəreɪt] *v.* 整合，合并

mathematical models 数字模型

algorithm [ˈælgərɪðəm] *n.* (尤指计算机)算法，运算法则

subsidiary [səbˈsɪdɪəri] *n.* 子公司

artificial intelligence 人工智能

Notes

1. In a bid to improve its forecasts, some outfits have started crowdsourcing data, including Weathernews, a firm in Chiba near Tokyo.

   Analysis："In a bid to do" 意为 "为了做某事"。"outfit" 意为 "协同工作的一组人，一队人"（尤指乐队、运动队、公司）。"crowdsource" 意为 "众包"，这一概念是由美国 Wired 杂志的记者 Jeff Howe 在 2006 年 6 月提出的；他给出的定义为一个公司或机构把过去由员工执行的工作任务；以自由自愿的形式外包给非特定的（而且通常是大型的）大众网络的做法。

   Translation：为提高预测准确性，一些机构开始通过众包的方式收集数据，比如，东京附近千叶市的天气新闻公司。

2. The company launched what they call the "Sakura Project" in 2004, signing up members who choose their own cherry tree and send pictures of its buds to the firm at regular intervals.

   Analysis："launch" 意为 "发动、发起（某项行动）"。

   Translation：这家公司在 2004 年推出了 "樱花项目"，与民众签约。民众选择所要观测的樱花树，并定期将花蕾照片发送给公司。

3. It also incorporates weather data collected from its own observation devices across Japan-13,000 locations in total, 10 times more than the official weather agency has.

   Analysis："observation devices" 意为 "观测设备"。

   Translation：公司还整合了用自己的观测设备收集的天气数据，这些设备遍布日本 1.3 万处观测场所，比日本气象厅的观测场所多 10 倍。

 Exercises

Translate the following short passages into Chinese.

1. As spring approaches in Japan, the country's weather forecasters face one of their biggest missions of the year: predicting exactly when the cherry blossoms will bloom. Japan's Sakura or cherry blossom season is feverishly anticipated by locals and visitors alike. Many tourists plan their entire trips around the blooms, and Japanese flock to parks in their millions to enjoy the seasonal spectacle.

_____

_____

_____

2. The company and other forecasters also employ mathematical models and algorithms. Otenki Japan, a forecaster run by a subsidiary of precision-equipment manufacturer Shimadzu, even began using artificial intelligence to predict cherry blossoms in 2018.

_____

_____

_____

3. The forecasts are not only for flower fans but reflect the fact that Sakura season is big business in Japan. Cherry blossoms symbolize the fragility of life in Japanese culture as full blooms only last about a week before the petals start falling off trees.

_____

_____

_____

Section III Grammar

Infinitive（不定式）

动词不定式的结构为 "to + 动词原形"，但有时也不带 to。其中 to 不是介词，而是动词不定式的符号，其否定结构是在 to 前面加上 not，及即 "not to + 动词原形"。动词不定式没有人称和数的变化，在句中不能做谓语。动词不定式和其后面的名词等构成不定式短语，在句子中可以用作主语、表语、宾语、补足语、定语、状语等。

一、不定式的构成及用法

不定式由 to + 动词原形构成。

不定式在句中不能作谓语。它具有动词的性质，本身可以带宾语和状语。动词不定式在句中除了不作谓语，可以作句子的任何成分。动词不定式的被动形式除了一般形式外，还有完成式和进行式。

动词不定式及其短语还可以有自己的宾语、状语，虽然动词不定式在语法上没有表面上的直接主语，但它表达的意义是动作，这一动作一定由施动者发出。动作的施动者我们称之为不定式的逻辑主语，其形式如下：

	主动形式	被动形式
一般式	（not）to make	（not）to be made
完成式	（not）to have made	（not）to have been made
进行式	（not）to be making	（not）to have been making

1. 语态

如果动词不定式的逻辑主语是这个不定式所表示的动作的承受者，不定式一般要用被动语态形式。如：

It's a great honor to be invited to Mary's birthday party.（不定式作主语）

It was impossible for lost time to be made up.（不定式作主语）

I wish to be sent to work in the country.（不定式作宾语）

Can you tell me which is the car to be repaired?（不定式作定语）

He went to the hospital to be examined.（不定式作状语）

在 there be 结构中，修饰主语的不定式可用被动，也可用主动。如：

There are still many things to take care of (to be taken care of).

但有时两种形式表达的意思不同，如：

There is nothing to do now.（We have nothing to do now.）

There is nothing to be done now.（We can do nothing now.）

2．时态

1）现在时：一般现在时表示的动词，有时与谓语动词表示的动作同时发生，有时发生在谓语动词的动作之后。

He seems to know this.

I hope to see you again. = I hope that I'll see you again. 我希望再见到你。

2）完成时：表示的动作发生在谓语动词表示的动作之前。

I'm sorry to have given you so much trouble.

He seems to have caught a cold.

3）进行时：表示动作正在进行，与谓语动词表示的动作同时发生。

He seems to be eating something.

4）完成进行时：表示动作在谓语动词之前发生，并且一直在进行着。

She is known to have been wreaking on the problem for many years.

二、不定式在句子中可以充当的成分

动词不定式在句中可充当主语、表语、宾语、宾补、定语和状语等。

1．作主语

动词不定式作主语可位于句首。例如：

To learn a skill is very important for everyone in society.

也可使用 it 作形式主语，而将其置于句末。例如：

It is necessary for young students to learn a foreign language.

2．作表语

动词不定式作表语常用于以下结构：My wish/job/aim/goal is…及 The next step/measure is…等。例如：

Your job is to type the papers in the office.

The next measure is to stop the river from being polluted.

3．作宾语

常见的只能使用动词不定式作宾语的动词有：agree，choose，decide，hope，fail，wish，refuse，expect，manage，plan，intend，pretend，promise，offer，afford，demand，arrange 等。例如：

They decided to build a highway between these two cities.

She offered to help me when I was in trouble.

4．作宾补

可后接动词不定式作宾补的动词有：allow，ask，advise，beg，cause，drive（强迫），encourage，expect，forbid，force，get，would like（love，hate），order，permit，persuade，teach，tell，want，warn，wish 等。例如：

The doctor advised her not to eat too much sugar.

I wish you to go to the meeting with me.

注：hope，demand，suggest 等动词后不能接动词不定式作宾补。

5. 作定语

动词不定式作定语，应位于所修饰词语之后，即：作后置定语。例如：

Have you got anything to eat? (to eat 修饰 anything，位于其后)

下列名词后常接动词不定式作定语：ability, attempt, chance, courage, decision, effort, failure, promise, way, wish 等。例如：

But she gave up the chance to go abroad.

由 only, first, last, next 以及序数词或形容词最高级修饰的名词后，也常接不定式作定语。例如：

Who was the last one to leave the classroom last night?

6. 作状语

动词不定式作状语，可表示目的、原因、结果或条件。例如：

We went there to see our grandparents. （目的）

I am very sorry to hear that. （原因）

She hurried home only to find her father dead. （结果）

To look at the picture, you would like it. （条件）

Exercises

Choose the best answer.

1. How kind you are! You always do what you can _____ others.

    A. help   B. helping   C. helps   D. to help

2. —Excuse me. Could you tell me _____ get to the nearest post office?

    —Sorry, I am new here.

    A. how can I   B. how I could   C. how to   D. what I can

3. Jessica's parents always encourage her _____ out her opinions.

    A. speak   B. speaking   C. to speak   D. will speak

4. The news reporters hurried to the airport, only _____ the film stars had left.

    A. to tell   B. to be told   C. telling   D. told

5. After receiving the Oscar for Best Supporting Actress, Anne Benedict went on _____ all the people who had helped in her career.

    A. to thank   B. thanking   C. having thanked   D. to have thanked

6. _____ more about Chinese culture, Jack has decided to take Chinese folk music as an elective course.

    A. Learn   B. Learned   C. To learn   D. To be learning

7. Sometimes I act as listening ear for fellow students _____ what is bothering them.

    A. to talk over   B. talked over   C. talk over   D. having talked over

8. _____ the early flight, we ordered a taxi in advance and got up very early.

    A. Catching   B. Caught   C. To catch   D. Catch

9. No matter how bright a talker you are, there are times when it's better _____ silent.

   A. remain   B. be remaining   C. having remained   D. to remain

10. I remembered _____ the door before I left the office, but forgot to turn off the lights.

   A. locking   B. to lock   C. having locked   D. to have locked

Section IV Writing

Letter of Apology（道歉信）

道歉信通常是因为自己的过失或疏忽，给别人带来了麻烦或损失，发觉后为了向对方道歉并请求对方谅解以增进友谊和信任的信函。道歉信的核心是向收信人表明歉意，请求对方谅解自己的过失或给对方带来的不快。写道歉信时要明确不是找借口为自己辩护，而是承认自己的过错并提出弥补过错的具体建议和方法，道歉的语言应该简练诚恳，并且希望收信人能够原谅自己的过失或疏忽。

Part 1 Sample

The following is a letter of apology. Please read and try to understand it.

Dear Mr. Davis，

    Much to my regret I was unable to keep my engagement to meet you at the park gate. I fear you are displeased at my failing to keep my promise, but I trust you will forgive me. Let me explain. My mother suddenly fell sick early yesterday morning, and I had to send her to hospital.

    I shall be obliged if you will kindly write and tell me when and where we may meet again. I hope to see you soon.

Yours sincerely,

Gina Evens

Part 2 Template

从上面的样例可以看出英语道歉信的书写结构为：

首先是称呼（Salutation）：当接受道歉的人是写信人熟悉的人时，称呼一栏一般用 Dear 加上名字，如 Dear John，Dear Bob 等，当接受道歉的人是长辈时，称呼一栏用 Dear Mr. / Miss/Ms. /Mrs. 加上收信人姓氏，如 Dear Mr. Liang，Dear Miss Taylor 等，如果无法确定收信人，称呼一栏可用 Dear 加上对方职位，如 Dear Sales Manager，Dear President 等或用 Dear Sir/Madam，或 To Whom It May Concern。

接下来为信的主体（Body）：道歉信的主体通常包含道歉的具体原因，应尽量详细描述事情经过，表示歉意时态度要真诚。结尾部分应再次表明承认错误的态度，请求收信人的谅解。也可以提出补救的办法。

最后是结尾（Conclusion）：一般写上 Yours truly，或 Yours sincerely，然后在下一行签名。

道歉信应及时之外，还必须写得诚恳，歉意应发自内心，不可敷衍塞责。再则，事情原委要解释清楚，措辞应当委婉贴切。

Dear _____（对象），

　　I am truly sorry that _____（写明道歉的事情）. The reason is that _____（解释事情发生的原因，消除误会和矛盾）. Hope you can accept my apologies and understand my situation. Is it possible _____（表明想补救的愿望，提出建议或安排）? Once again, I am sorry for _____（再次表示遗憾和歉意）.

Yours Sincerely,
Liu Wei

Part 3　Useful Patterns

1. I am writing to apologize for…/I am writing to say sorry for…

2. I would like to give you my apology for…

3. I am really sorry for doing…

4. I feel truly sorry about it and want you to know what happened.

5. Please accept my apologies for the oversight.

6. The reason why…was that…

7. Once again, I'm sorry for any inconvenience caused.

8. I sincerely hope you can accept my apologies and understand my situation.

9. Let's not allow a little misunderstanding to come between us.

10. I am indeed very sorry for what I said/did, but it was not my intention to cause offence.

1. 我写这封信是因…向你致歉。

2. 对于…我向你道歉。

3. 为做了…事我深表歉意。

4. 我为此感到非常内疚，所以想让您知道实情。

5. 请原谅我的疏忽。

6. 某事发生的原因是…

7. 我再一次为造成的不便表示歉意。

8. 我真心希望你能理解我的处境，并接受我的道歉。

9. 可别让我们之间存在哪怕一点点的误解。

10. 对于我说的话/做过的事，我确实感到很抱歉。但请相信，我并非要故意冒犯你。

Part 4　Exercises

Ⅰ. Translate the following letter of apology into Chinese.

Dear Amanda,

　　I am terribly sorry to tell you that I am unable to attend your birthday party next Thursday evening. That is owing to the fact that my younger brother suddenly fell ill and was taken to a hospital this morning. I have to go there immediately and take care of him. As told by the doctor in charge, it will take around five days for him to recover and I have asked my boss for a leave.

　　I really regret that I cannot go to celebrate your birthday personally and would miss the perfect chance of enjoying myself with all our old friends. I have chosen a small gift for you and will send it to you tomorrow to show my best wishes. Besides, please give my regards to our friends when you meet them at the party.

Yours sincerely,
Li Lei

Ⅱ. Write a letter of apology according to the information given in Chinese.

AmyDiamond 因为学院要开会而无法脱身，错过了和 Ken 的约会，请以她的名义写一封道歉信。Amy 由于失约，提议再次安排见面，告知 Ken 自己一周内除了周五下午 2 点到 4 点没空外，其他时间都可以；也可以安排在办公室见面。

注意：不要逐句翻译，字数 100 个单词左右。

Unit 7

Artificial Intelligence

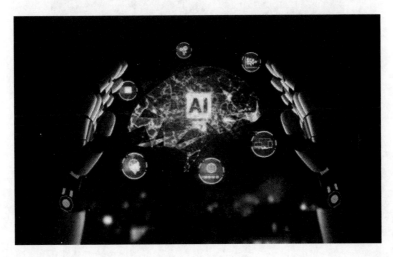

Unit Goals

After learning this unit, you will be able to:

- talk about artificial intelligence
- understand the passages and grasp the key words and expressions
- review the grammar of gerund
- write a letter of inquiry

Section I Listening & Speaking

myriad 极大数量

implement 实施；执行

interact 相互作用；相互联系

career 生涯；职业

promising 有希望的；有前途的

option 选择权；可选物

rapid 急速的；迅速的

integration 集成；综合

logistics 物流；后勤学

artificial 人工的

Artificial Intelligence 人工智能

emerge 出现

imitation 模仿；效法

synthesis 合成

solid 固体的；可靠的

bound 必定的

specific 明确的；特殊的

recognition 识别；认出

domain 领地；领域

render 提供；使成为

intelligence 智能

Read the conversations carefully and then complete the communicative tasks.

Conversation 1

A：Myriads of AI techniques have **emerged** in the past decade for **implementing** and building AI systems.

B： I am not sure that I have heard any. Can you name one for me?

A： There are several typical ones, including Natural Language Processing, Neural Networks, and Vector Machines.

B： I only seemed to have heard the first one.

A： You mean Natural Language Processing?

B: Right, also known as NPL.

A: Correct! That means you understand what it means!

B: Not really. Is it a kind of **imitation** to human brain?

A: That is close. In a one – liner, Natural Language Processing is the study of how a computer **interacts** with a human language.

B: So with this technique, we can carry on speech recognition and speech **synthesis** in human language?

A: Bingo, this technique is already in application phase and companies are using it in their voice assistants.

B: Oh, I know, like Apple's Siri, Google Assistant, and Microsoft's Cortana.

A: You do know about this field.

Key Words

myriad   极大数量

emerge   出现

implement   实施；执行

imitation   模仿；效法

interact   相互作用；相互联系

synthesis   合成

Task 1   Act out the conversation with your partner based on the following clues.

A: Myriads of AI techniques have emerged in the past decade for implementing and building AI systems.

B: I am not sure that I have heard any. _____①_____?

A: There are several typical ones, including Natural Language Processing, Neural Networks, and Vector machines.

B: _____②_____.

B: Right, also known as NPL.

A: Correct! _____③_____!

B: Not really. _____④_____?

A: That is close. In a one-liner, Natural Language Processing is the study of how a computer interacts with a human language.

B: So with this technique, we can carry on speech recognition and speech synthesis in human language?

A: Bingo, this technique is already in application phase and companies are using it in their voice assistants.

B：Oh, I know, like Apple's Siri, Google Assistant, and Microsoft's Cortana.

A：_____⑤_____.

Conversation 2

A：Do you have any idea about your **career** direction?

B：I have been considering about AI related fields.

A：That's a **solid** plan. AI related jobs have **promising** future.

B：That's also what I deem.

A：But you still seem confused about your choice.

B：Right, because I don't know the **specific** segment I should choose.

A：Don't worry. Rise of AI has given way to new technologies which are **bound** to change the way we live our lives.

B：Really? What are they?

A：They are：Cloud Computing, Internet of Things, Big Data, and Open Source Technologies.

B：Great, I have never imagined so many **options**.

A：But actually, if you want to develop your career or create an outstanding product, you'd better just choose one specific field of AI rather than trying to cover them all.

B：That's wise. I will choose Big Data as a direction, which is closest to my major.

Key Words

career 生涯；职业

solid 固体的；可靠的

promising 有希望的；有前途的

specific 明确的；特殊的

bound 必定的

option 选择权；可选物

Task 2 Act out the conversation with your partner based on the following clues.

A：Do you have any idea about your career direction?

B：_____①_____.

A：That's a solid plan. _____②_____.

B：That's also what I deem.

A：_____③_____.

B：Right, because I don't know the specific segment I should choose.

A：Don't worry. Rise of AI has given way to five new technologies which are bound to change the way we live our lives.

B：Really? What are they?

A：They are：Cloud Computing, Internet of Things, Big Data, and Open Source Technologies.

B：Great, _____④_____.

A：But actually, if you want to develop your career or create an outstanding product, you'd better just choose one specific field of AI rather than trying to cover them all.

B：That's wise. _____⑤_____, which is closest to my major.

Conversation 3

A：Are you familiar with the development of AI technology?

B：It must be **rapid**, right?

A：You bet. In the last five years, the field of AI has made major progress in almost all its standard sub-areas.

B：What are those sub-areas?

A：Including vision, speech **recognition** and generation, natural language processing (understanding and generation), image and video generation, decision-making, and **integration** of vision and motor control for robotics.

B：So, that means there must be numerous breakthroughs in all fields.

A：Yes. Breakthrough applications emerged in a variety of **domains** including games, medical diagnosis, **logistics** systems.

B：Those signs are also obvious in autonomous driving, language translation, and interactive personal assistance.

A：The core technology behind most of these most visible advances is machine learning, especially deep learning.

B：Yes, we should appreciate technology development **rendering** us a better life.

 Key Words

rapid　急速的；迅速的
recognition　识别；认出
integration　集成；综合
domain　领地；领域
logistics　物流；后勤学
render　提供；使成为

Task 3　Act out the conversation with your partner based on the following clues.

A：Are you familiar with the development of AI technology?

B：_____①_____?

A: You bet. In the last five years, the field of AI has made major progress in almost all its standard sub-areas.

B: _____②_____?

A: Including vision, speech recognition and generation, natural language processing (understanding and generation), _____③_____, decision-making, and integration of vision and motor control for robotics.

B: So, _____④_____.

A: Yes. Breakthrough applications emerged in a variety of domains including games, medical diagnosis, logistics systems.

B: Those signs are also obvious in autonomous driving, language translation, and interactive personal assistance.

A: The core technology behind most of these most visible advances is machine learning, especially deep learning.

B: Yes, _____⑤_____.

听力材料

Part C Passages

Listen to the following passages carefully and fill in the blanks with the information you've heard.

Passage 1

Nowadays, AI is used in almost all _____, giving a technological _____ to all companies integrating AI at scale. According to McKinsey, AI has the _____ to create 600 billion dollars of value in _____ and to bring 50 per cent more incremental value in banking compared with other analytics techniques. In _____ and logistics, the potential revenue jump is 89% more.

Passage 2

In a _____, AI provides cutting-edge technology to deal with _____ data that a human being cannot handle. AI _____ redundant jobs, allowing a worker to focus on the high level, value-added tasks. When AI is _____ at scale, it leads to cost reduction and _____ increase.

Passage 3

Artificial Intelligence improves an _____ product. Before the age of machine learning, core products were built upon _____ rules. Firms introduced Artificial Intelligence to _____ the functionality of the product rather than starting from _____ to design new products. You can think of a Facebook image. A few years ago, you had to tag your friends _____. Nowadays, with the help of AI, Facebook gives you a friend's recommendation.

Passage 4

Machine learning is an _____ field, meaning it needs data to test new ideas or _____ . With the _____ of the internet, data became more easily _____ . Besides, _____ companies like NVIDIA and AMD have developed high-performance graphics chips for the gaming market.

Passage 5

That means, AI is not _____ related to computer science. This is a _____ of study that encompasses human behavior, _____ , psychology, and even language and linguistics. There's still not a common _____ among academicians about its _____ .

Section Ⅱ   Reading

Passage 1

AI Introduction

AI (Artificial Intelligence) is a machine's ability to perform **cognitive functions** as humans do, such as **perceiving**, learning, reasoning, and solving problems. The **benchmark** for AI is the human level concerning in terms of reasoning, speech, and vision. [①]

Nowadays, AI is used in almost all industries, giving a technological edge to all companies **integrating** AI at scale. [②] According to McKinsey, AI has the potential to create 600 billion dollars of value in retail and to bring 50 per cent more **incremental** value in banking compared with other **analytics** techniques. In transport and logistics, the potential **revenue** jump is 89% more.

Concretely, if an organization uses AI for its marketing team, it can automate **mundane** and repetitive tasks, allowing the sales representative to focus on relationship building, lead **nurturing**, etc. [③] A company named Gong provides a conversation intelligence service. Each time a Sales Representative makes a phone call, the machine records, transcribes and analyzes the chat. [④] The VP can use AI analytics and recommendation to **formulate** a winning strategy.

In a nutshell, AI provides cutting-edge technology to deal with complex data that a human being cannot handle. AI automates **redundant** jobs, allowing a worker to focus on the high level, value-added tasks. When AI is implemented at scale, it leads to cost reduction and revenue increase. [⑤]

 New Words and Expressions

cognitive [ˈkɒɡnətɪv]   adj. 认识的；认知的
function [ˈfʌŋkʃn]   n. 职务；功能
perceive [pəˈsiːv]   v. 注意；觉察；理解
benchmark [ˈbentʃmɑːk]   n. 基准点；参照点
integrate [ˈɪntɪɡreɪt]   v. 整合；结合
incremental [ˌɪŋkrəˈmentl]   adj. 增加的；增量的

analytics [ˌænəˈlɪtɪks]　*n.* 分析学；解析学

revenue [ˈrevənuː]　*n.* 税收；收入

mundane [mʌnˈdeɪn]　*adj.* 平凡的；世俗的

nurture [ˈnɜːtʃə(r)]　*v.* 养育；培育；照顾

formulate [ˈfɔːmjuleɪt]　*v.* 规划；制订

redundant [rɪˈdʌndənt]　*adj.* 多余的，过多的

Notes

1. The benchmark for AI is the human level concerning in terms of reasoning, speech, and vision.

   Translation：人工智能的基准是达到人类在推理、言语和视觉方面的思考水平。

2. Nowadays, AI is used in almost all industries, giving a technological edge to all companies integrating AI at scale.

   Analysis："giving a technological edge…" 此处现在分词表示伴随动作，是指同时发生的影响。

   Translation：如今，人工智能几乎应用于所有行业，为所有大规模集成人工智能的公司提供了技术优势。

3. Concretely, if an organization uses AI for its marketing team, it can automate mundane and repetitive tasks, allowing the sales representative to focus on relationship building, lead nurturing, etc.

   Analysis：本句中使用了"if"引导的条件状语从句，而在主句中，"allowing the sales representative to…" 使用现在分词表示动作的结果。"lead" 此处是商业术语，表示"在销售过程中获得的信息线索"。

   Translation：具体地说，如果一个组织将人工智能用于其营销团队，它可以自动化单调和重复的任务，让销售代表专注于建立关系、培养潜在客户等。

4. Each time a Sales Representative makes a phone call, the machine records, transcribes and analyzes the chat.

   Analysis："each time" 作为一个整体，引导了时间状语从句，表示"每一次"。在主句部分，三个动词并列，最后一个动词前用"and"连接。

   Translation：每次销售代表打电话，机器都会记录、转录和分析聊天内容。

5. When AI is implemented at scale, it leads to cost reduction and revenue increase.

   Translation：当人工智能被大规模实施时，就会带来成本降低和收入增加。

Exercises

Ⅰ. Answer the following questions according to the text.

1. What is Artificial Intelligence?

_____

_____

2. What is the benchmark for AI?

_____

_____

3. According to McKinsey, what potential does AI have in banking?

_____

_____

4. If an organization uses AI for its marketing team, what kinds of work can it do?

_____

_____

5. When AI is implemented at scale, what can it lead to?

_____

_____

II. Fill in the blanks with words according to the meaning of the article by memory.

_____, if an organization uses AI for its _____ team, it can automate _____ and repetitive tasks, allowing the sales _____ to focus on relationship building, lead _____, etc. A company named Gong provides a conversation intelligence service. Each time a Sales Representative makes a phone call, the machine records, _____ and analyzes the chat. The VP can use AI analytics and recommendation to _____ a winning strategy.

Concretely marketing mundane represe ntativenurturing transcribes formulate

III. Fill in the blanks with the words given below. Change the forms when necessary.

perceive	incremental	revenue
mundane	cognitive	nurture
benchmark	redundant	

1. Concretely, if an organization uses Artificial Intelligence for its marketing team, it can automate _____ and repetitive tasks.

2. AI is a machine's ability to perform _____ functions as humans do.

3. We can _____ his sorrow by the looks on his face.

4. AI automates _____ jobs, allowing a worker to focus on the high level, value-added tasks.

5. According to McKinsey, AI has the potential to create 600 billion dollars of value in retail

and to bring 50 per cent more _____ value in banking compared with other analytics techniques.

6. In transport and logistics, the potential _____ jump is 89% more.

7. The _____ for AI is the human level concerning in terms of reasoning, speech, and vision.

8. She _____ the child as if he had been her own.

IV. Translate the following sentences into Chinese.

1. The benchmark for AI is the human level concerning in terms of reasoning, speech, and vision.

2. Nowadays, AI is used in almost all industries, giving a technological edge to all companies integrating AI at scale.

3. Each time a Sales Representative makes a phone call, the machine records, transcribes and analyzes the chat.

4. The VP can use AI analytics and recommendation to formulate a winning strategy.

5. AI automates redundant jobs, allowing a worker to focus on the high level, value-added tasks.

6. According to McKinsey, AI has the potential to create 600 billion dollars ofvalue in retail and to bring 50 per cent more incremental value in banking compared with other analytics techniques.

Passage 2

Where is AI Used?

AI has broad applications.

Artificial Intelligence is used to **reduce** or avoid repetitive tasks. **For instance**, AI can repeat a task continuously, without fatigue. AI never rests, and it is **indifferent** to the task to carry out. [1]

Artificial Intelligence improves an existing product. Before the age of machine learning, core products were built upon hard-code rules. Firms introduced artificial intelligence to **enhance** the functionality of the product rather than **starting from scratch** to design new products. [2] You can think of a Facebook image. A few years ago, you had to tag your friends **manually**. Nowadays, with the help of AI, Facebook gives you a friend's **recommendation**.

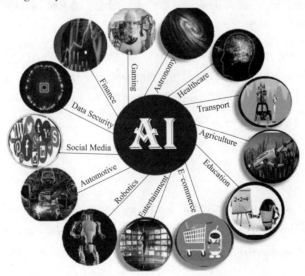

AI is used in all industries, from marketing to supply chain, finance, food-processing sector. According to a McKinsey survey, financial services and hightech communication are leading the AI fields.

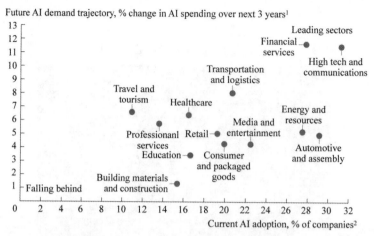

Why is AI booming now? Let's understand it by the **diagram** below.

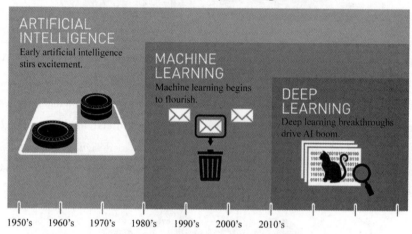

Since an early flush of optimism in the 1950s, smaller subsets of artificial intelligence - first machine learning, then deep learning, a subset of machine learning - have created ever larger disruptions.

A neural network has been out since the nineties with the seminal paper of Yann LeCun. However, it started to become famous around the year 2012. Explained by three critical factors for its popularity:

1. Hardware

2. Data

3. **Algorithm**

Machine learning is an **experimental** field, meaning it needs data to test new ideas or approaches. [3] With the boom of the Internet, data became more easily **accessible**. [4] Besides, giant companies like NVIDIA and AMD have developed high-performance **graphics** chips for the gaming market.

New Words and Expressions

reduce [rɪˈdjuːs]　*v.* 减少；缩小

for instance　例如；以……为例

indifferent [ɪn'dɪfrənt] *adj.* 漠不关心的；中立的

enhance [ɪn'hæns] *v.* 提高；增加；加强

start from scratch 从头做起；白手起家

manually ['mænjuəli] *adv.* 手动地；手工地

recommendation [ˌrekəmen'deɪʃn] *n.* 推荐；建议

diagram ['daɪəgræm] *n.* 图解；图表

algorithm ['ælgərɪðəm] *n.* 算法

experimental [ɪkˌsperɪ'mentl] *adj.* 实验的；实验性的

accessible [ək'sesəbl] *adj.* 可得到的；可进入的

graphics ['græfɪks] *n.* 制图法；绘图学

Notes

1. AI never rests, and it is indifferent to the task to carry out.

   Analysis：此处"indifferent"一定避免误译成"相同"。

   Translation：人工智能从不休息，它对要执行的任务一视同仁。

2. Firms introduced artificial intelligence to enhance the functionality of the product rather than starting from scratch to design new products.

   Analysis："rather than"表示"而不是"。

   Translation：公司引入人工智能来增强产品的功能，而不是从头开始设计新产品。

3. Machine learning is an experimental field, meaning it needs data to test new ideas or approaches.

   Translation：机器学习是一个实验领域，这意味着它需要数据来测试新的想法或方法。

4. With the boom of the Internet, data became more easily accessible.

   Translation：随着互联网的蓬勃发展，数据变得更加容易获取。

Exercises

Answer the following questions according to the text.

1. What is Artificial Intelligence used to do?

   _____

   _____

2. What attitude does AI have towards the task to carry out?

   _____

   _____

3. Before the age of machine learning, what were core products built upon?

4. What industries is AI used in? Can you name several examples?

5. What three critical factors can explain AI's booming popularity?

Passage 3

A Brief Introduction to Artificial Intelligence

AI is the new **buzzword** of the 21st century. Artificial Intelligence is coming in to our lives faster than we had **anticipated**. [1] It's helping us in shopping. It's at the other end when you're talking to customer service centre. It's driving our cars and even recommending you videos on YouTube. It's making sure that it knows about you more than you do.

Alan Turing defined AI as:

"Artificial Intelligence is the science of making machines do things that would require intelligence if done by man." [2]

Alan Turing / 1912 - 1954

That means, AI is not specifically related to computer science. This is a field of study that **encompasses** human behavior, biology, psychology, and even language and linguistics. There's still not a common **consensus** among academicians about its definition.

In thisarticle, we try to give a broader picture of AI. How it is organized and its **various** areas and fields of study. First we will discuss the **terminologies associated** with AI and then we will discuss the techniques used in **implementing** AI.

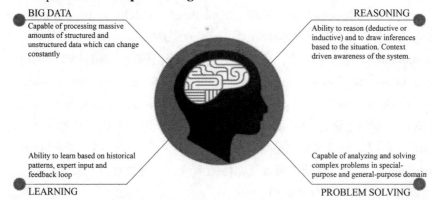

BIG DATA
Capable of processing massive amounts of structured and unstructured data which can change constantly

REASONING
Ability to reason (deductive or inductive) and to draw inferences based to the situation. Context driven awareness of the system.

Ability to learn based on historical patterns, expert input and feedback loop

LEARNING

Capable of analyzing and solving complex problems in special-purpose and general-purpose domain

PROBLEM SOLVING

One of the key AI terminologies is Machine Learning.

First thing to understand about machine learning is that it is an application of AI. It is the process by which we create systems that have ability to learn with experience. [3]

For instance, the systems which automatically **identify spam emails** are trained to do so by exposing them to millions of e-mails which are spammy and non-spammy in nature. [4] With more data, the program is able to understand and learn what makes an e-mail.

New Words and Expressions

buzzword [ˈbʌzwɜːd]   *n.* 流行词；行话

anticipate [ænˈtɪsɪpeɪt]   *v.* 预料，预测

encompass [ɪnˈkʌmpəs]   *v.* 包围；包括

consensus [kənˈsensəs]   *n.* 一致；同意；共识

various [ˈveərɪəs]   *adj.* 各种各样的

terminology [ˌtɜːmɪˈnɒlədʒi]   *n.* 术语；术语学

associated [əˈsəʊʃieɪtɪd]   *adj.* 相关的；联合的

implement [ˈɪmplɪmənt]   *v.* 实施；执行

identify [aɪˈdentɪfaɪ]   *v.* 鉴定；识别；辨认出

spam e-mail   垃圾邮件

Notes

1. Artificial Intelligence is coming in to our lives faster than we had anticipated.
   Analysis：本句含有"than"引导的比较状语从句。

Translation：人工智能走进我们生活的速度比我们预期得更快。

2. Artificial Intelligence is the science of making machines do things that would require intelligence if done by man.

Analysis：本句"things"后接限制性定语从句，而定语从句中使用了"if"引导的条件状语从句。

Translation：人工智能是一门让机器做一些需要人类智能的事情的科学。

3. It is the process by which we create systems that have ability to learn with experience.

Analysis：本句含有两个定语从句，先行词分别是"process"和"systems"。

Translation：它是一个过程，通过这个过程，我们创建了具有经验学习能力的系统。

4. For instance, the systems which automatically identify spam emails are trained to do so by exposing them to millions of e-mails which are spammy and non-spammy in nature.

Analysis：本句主语是"the systems"，后接定语从句。本句谓语部分为"are trained to do so"。所以本句的主干就是"系统被训练做……"。其余部分则为修饰和说明。

Translation：例如，自动识别垃圾邮件的系统，通过数百万封垃圾邮件和非垃圾邮件的训练后，才做到了自动识别。

Exercises

Translate the following short passages into Chinese.

1. AI is the new buzzword of the 21st century. Artificial Intelligence is coming in to our lives faster than we had anticipated.

2. Artificial Intelligence is the science of making machines do things that would require intelligence if done by man.

3. That means, AI is not specifically related to computer science. This is a field of study that encompasses human behaviour, biology, psychology, and even language and linguistics.

Section Ⅲ　Grammar

Gerund(动名词)

非谓语动词是动词的一种形式，它不作为主语或谓语，也不表达时态和语态，通常在句子中作名词、形容词或副词的补语、定语或状语。常见的非谓语动词包括不定式(infinitive)、动名词(gerund)和分词(participle)。

动名词(gerund)是一个非谓语动词形式，它以-ing 结尾，常常用作名词。它可以单独作为主语、宾语、表语、介词宾语或复合结构的组成部分。同时，动名词仍然具有动词的若干特点，它可以有自己的宾语和状语。

一、动名词在句中所能充当的成分

	主语	宾语	表语	定语	状语	补语
动名词	√	√	√			

1. 动名词作主语

Smoking is harmful to your health.

(吸烟对健康有害。)

Writing essays about literature is challenging but rewarding.

(写有关文学的论文既具有挑战性又有回报。)

注意：动名词作主语时，谓语动词一律用单数。

A. 用 it 作形式主语

用动名词作主语时，和动词不定式一样，有时也用 it 作形式主语。

It is no use (useless) talking too much.　=　Talking too much is no use (useless).

It trains the ear listening to music.　=　Listening to music trains the ear.

B. 在 there be 句型中

只能用动名词作主语(不用不定式)

There is no + ing. (…是不可能的)

　= It is impossible to…

　= No one can…

There is no foreseeing what will happen.

　= It is impossible toforesee what will happen.

　= No one can foresee what will happen.

无法预言将会发生什么事情。

There is no denying the importance of education.

　= It is difficult to deny the importance of education.

教育的重要性是不可否认的。

2. 动名词作表语

His favorite activity is playing basketball.

（他最喜欢的活动是打篮球。）

His biggest fear is not achieving his goals.

（他最大的恐惧是无法实现自己的目标。）

动名词作表语时，一般表示比较抽象的习惯动作，表语和主语常常可以互换。

His hobby is painting.

Painting is his hobby.

What he likes best is making jokes.

Making jokes is what he likes best.

3. 动名词作宾语

I enjoy reading books before bedtime.

（睡前我喜欢读书。）

I have been considering studying abroad for a year.

（我一直在考虑出国留学一年。）

A. 只能用动名词作宾语的动词有

admit 承认, appreciate 感激、赞赏, avoid 避免, complete 完成, consider 考虑, delay 耽误, deny 否认, detest 讨厌, endure 忍受, enjoy 喜欢, escape 逃脱, prevent 阻止, fancy 想象, finish 完成, imagine 想象, mind 介意, miss 错过, postpone 推迟, practise 训练, recall 回忆, resent 讨厌, resist 抵抗, resume 继续, risk 冒险, suggest 建议, face 面对, include 包括, stand 忍受, understand 理解, forgive 宽恕, keep 继续

B. 只能用动名词作宾语的词组有

admit to 承认, add to 加上, be（get）used to 习惯于, be accustomed to 习惯于, be tired of 厌烦…, be fond of 喜欢, be capable of 有…的能力, be afraid of 害怕…, be proud of 为…自豪、骄傲, be successful in 在…成功, be good at 擅长于…, be interested in 对….感兴趣, be engaged in 从事于, be busy with 忙于, be sentenced to 宣判, burst out 突然, can't help 禁不住, come to 谈到, confess to 承认, count on/upon 依靠、信赖, devote oneself to 把…贡献给…, dream of 梦想, depend on 依靠, excuse…for…为…而原谅, feel like 愿意做什么, give up 放弃, hold off 推迟, insist on 坚持, It's worth…值得, It's no use/good 没有用/不好 keep on 坚持, know of 对…了解, look forward to 期盼、盼望, lead to 导致, object to 反对, put off 推迟, prevent…from…阻止…免受…, pay attention to 注意, prefer…to 比起…更…, stick to 坚持, set about 着手去做, think of/about 想到/考虑

二、动名词的时态和语态

1. 动名词的时态

A. 动名词的一般式

表示泛指的一般性动作，即动作发生的时间并不明确；或者动名词所表示的动作与谓语动词的动作同时发生或在它的动作之后发生。

She loves cooking all kinds of food, from Chinese to Italian.

（她喜欢烹饪各种食物，从中餐到意大利菜。）

He needs to start exercising more regularly to improve his health.

（他需要开始更加定期地锻炼以改善他的健康状况。）

B. 动名词的完成式

所表示的动作在谓语动词所表示的动作之前发生。

They have discussed starting a business together for a long time.

（他们已经讨论了很长时间一起创业的事情。）

They have been considering moving to a new city for better job opportunities.

（他们一直在考虑搬到一个新城市以寻求更好的工作机会。）

2. 动名词的语态

A. 动名词被动语态

当动名词逻辑上的主语是动名词所表示的动作的对象时，动名词一般要用它的被动式。

The squirrel was lucky that it just missed being caught.

（那只松鼠很幸运，它差点被捉住了。）

Before being used, the machine must be checked.

（机器在使用前必须进行检查。）

B. 动名词的主动形式代替被动形式

主语（通常是物）＋　　want

need

require（需要）＋doing（动名词的主动形式）

deserve（值得）

＝主语＋want/ need/ require/ deserve to be done（不定式的被动形式）

The radio wants (needs, requires) repairing.

＝The radio wants (needs, requires) to be repaired.

另外：

主语＋be worth doing　＝　主语＋be worthy＋to be done/ of being done.

The book is worth reading again.

＝The book is worthy to be read again.

＝The book is worthy of being read again.

Exercises

Choose the best answer.

1. When I started learning English, I focused on _____ lots of vocabulary.

　　A. remember　　　B. remembering　　C. remembered　　　D. having remembered

2. The time and effort he has devoted during the past few years _____ trees in that remote area is now considered to be of great value.

A.　to planting　　　　B.　to plant　　　　C.　plant　　　　　　D.　planting

3.　As a result of the earthquake, two thirds of the buildings in the area _____ .

A.　need repairing　　　　　　　　B.　needs repairing

C.　needs to be repaired　　　　　　D.　need to repair

4.　As long as you keep on _____ hard, you'll get promoted sooner or later.

A.　work　　　　B.　be working　　　　C.　worked　　　　D.　working

5.　Living in the dormitory means _____ to take care of yourself, understandeach other's differences and share _____ interests.

A.　to learn; common　　　　　　B.　learning; ordinary

C.　to learn; ordinary　　　　　　D.　learning; common

6.　Harvey finished _____ his research paper, but he needed to check it before submitting it to the professor.

A.　to type　　　　B.　typed　　　　C.　typing　　　　D.　type

7.　The sales department of a company is engaged in _____ the products and making profits.

A.　selling　　　　B.　sell　　　　C.　being sold　　　　D.　having sold

8.　A survey suggests that nearly one in six children has difficulty _____ to talk.

A.　to learn　　　　B.　learning　　　　C.　learn　　　　D.　learnt

9.　_____ well in an interview will be an important part of getting a place at university.

A.　Do　　　　B.　Doing　　　　C.　Done　　　　D.　Being done

10.　Sometimes _____ a business can feel like a tough decision to make, no matter how good your idea is.

A.　starting　　　　B.　being started　　　　C.　start　　　　D.　to be started

Section Ⅳ　Writing

Letter of Inquiry（咨询信）

咨询信（Letter of Inquiry）是一种用于商务和官方交流的书信，通常被用于向专业人士或组织请求建议、信息或意见。咨询信的主要目的是寻求所需的信息或帮助。

Part 1　Sample

The following is a letter of inquiry. Please read and try to understand it.

October 8th，2019 To whom it may concern， 　　I am writing to you to apply for the membership of the English Club. As a sophomore from the English Department，I love English literature very much. I often attend the English Corner held on Friday in the People's University of China and would like to discuss various topics in English with others. 　　I am eager to be one member of the English Club and improve my English. I know there are some requirements，but I am not clear about the details. Is it necessary for me to pay for membership? If so，how much should I pay? And how often? I have also heard that the club will host various activities，but can you give me more details? 　　I would be very grateful if you can send me a reply as soon as possible. Looking forward to hearing from you soon. 　　　　　　　　　　　　　　　　　　　　Yours sincerely， 　　　　　　　　　　　　　　　　　　　　　　Wang Mei	（1）开篇点名写信缘由 （2）说明自己的身份和对英语的爱好 （3）表明加入俱乐部的殷切期望和目的 （4）引出下文，提出问题 （5）结束语，期待对方的回信

Part 2　Template

从上面的样例可以看出，英语咨询信（Letter of Inquiry）与其他书信的书写规则基本相同，通常包括信头、称呼、正文、结尾和署名等部分，其写作要点如下：

首段：表明写作意图，说明需要咨询的具体事项或问题。

主体：详细说明需要咨询的内容，并向收信人提出相关问题。

尾段：向对方表示真诚的感激，邀请收信人就咨询的问题提供意见或建议，并提供联系方式。

Dec. 20, 2018

Dear _____ ,

I am _____ . I am writing to see if it is possible for you to provide me with _____ information regarding _____ .

First of all, what are _____ ? Secondly, when will _____ ? Thirdly, is _____ ?

I would also like to inquire _____ . Could you be so kind as to send me some relevant booklets on the above-mentioned aspects?

Thank you for your kindness, and your prompt attention to this letter will be highly appreciated.

Yours sincerely,

Li Ming

Part 3　Useful Patterns

1. I am writing to inquire about…

2. My name is… and I am planning to study at your university during the summer break.

3. I plan to start classes next semester and would greatly appreciate it if you could provide me with some necessary information.

4. I have obtained a Bachelor's degree in Biology and would like to continue my studies at your university.

5. Could you please provide me with some information about the … situation at your university?

6. Would it be possible for you to inform me of this information as soon as possible and mail me the relevant forms?

7. I apologize for any inconvenience and appreciate your kind assistance.

8. I would greatly appreciate your response.

1. 我写信是要询问…

2. 我是…，暑期计划到贵校学习。

3. 我计划下个学期开始上课，如果您能告知一些必要的信息，我将不胜感激。

4. 我已经取得了生物学学士学位，并希望能在贵校继续学习。

5. 您能向我介绍一下贵校的…情况吗？

6. 您能尽快告知这些信息并邮寄给我相关表格吗？

7. 很抱歉打扰您，对您的友善帮助不胜感激。

8. 如果您能回复，我将感激不尽。

I. Translate the following letter of inquiry into Chinese.

Dear Sir or Madam,

　　I am a Chinese student who wishes to pursue further study at your university. I plan to start my course next semester, and it would be highly appreciated if you would provide me with some essential information.

　　First, how much are the tuition fees? Though I intend to be self-supporting, I would be interested to know if there are any scholarships available for international students. Second, I wonder what qualifications I need to follow a course at your university, now I have got a master's degree in a Chinese university, is it enough? Third, as regards to accommodations, if possible, I would like to have a single room, but if single rooms are expensive, I would be willing to share.

　　I am looking forward to your reply, and to attending your esteemed institution.

Yours sincerely,

Li Hua

II. Write a letter of inquiry according the information given below.

　　Suppose you are a store holder and want to purchase goods for Christmas sales. Write to a wholesaler to ask the detailed information of the available goods. You should write about 100 words. Do not sign your own name at the end of the letter. Use "George O. Justin" instead.

Unit 8

Internet of Things

Unit Goals

After learning this unit, you will be able to:

- talk about Internet of Things
- understand the passages and grasp the key words and expressions
- review the grammar of participle
- write a tour commentary

Section I Listening & Speaking

Part A Words and Expressions

gurus 领袖；专家	speculation 推测
framework 框架；体系	allocate 分派；分配
scope 范围	domain 领地；领域
thrive 兴旺；繁荣	vulnerability 弱点；脆弱
deem 认为	device 装置；设备
hack 非法侵入	scenario 情形，情景
facility 设备；设施	breach 违反；在…上打开缺口
acronym 首字母缩略词	various 各种各样的
sector 部门；领域	debug 调试
conduct 处理；实施	advancement 前进；进步
hospitality 酒店管理	

Part B Conversations

Read the conversations carefully and then complete the communicative tasks.

Conversation 1

A: Do you know how fast the spread of IoT is?

B: I have heard that all market **gurus** highly favor the rise of IoT application development technology.

A: Yes, they see it as the next best thing in modern technology.

B: So, how high is **speculation** about the growth of the IoT space?

A：There are views that the IoT industry could grow to a market cap of almost ＄1. 4 trillion USD in the next 3 years.

B：Really? Is that data solid?

A：Certainly, as quoted by the sources in Business Insider, the number of devices that will be connected to the IoT **framework** could easily cross 24 billion devices.

B：Wow, that number is way over my imagination.

A：Yeah, that number is huge, even though it **allocates** a high percentage of business and government use.

B：That's the question. There is still a lot of **scope** for consumer services supported by IoT devices.

A：Yes, for example, even automobiles could be linked to the IoT **domain** and approximately a quarter billion vehicles could be connected to the Internet.

B：There is really some room for IoT to **thrive**.

A：Yes. It is estimated that China and the United States could be the largest markets for the IoT by the year 2024 with over 27 billion machine-to-machine linkages.

 Key Words

gurus　领袖；专家

speculation　推测

framework　框架；体系

allocate　分派；分配

scope　范围

domain　领地；领域

thrive　兴旺；繁荣

Task 1　Act out the conversation with your partner based on the following clues.

A：＿＿＿＿①＿＿＿＿?

B：I have heard that all market gurus highly favor the rise of IoT application development technology.

A：Yes, ＿＿＿＿②＿＿＿＿.

B：So, how high is speculation about the growth of the IoT space?

A：There are views that the IoT industry could grow to a market cap of almost ＄1. 4 trillion USD in the next 3 years.

B：Really? Is that data solid?

A：Certainly, as quoted by the sources in Business Insider, the number of devices that will be connected to the IoT framework ＿＿＿＿③＿＿＿＿.

B: Wow, that number is way over my imagination.

A: Yeah, that number is huge, even though it allocates a high percentage of business and government use.

B: That's the question. _____④_____ .

A: Yes, for example, even automobiles could be linked to the IoT domain and approximately a quarter billion vehicles could be connected to the Internet.

B: _____⑤_____ .

A: Yes. It is estimated that China and the United States could be the largest markets for the IoT by the year 2024 with over 27 billion machine-to-machine linkages.

Conversation 2

A: Have you ever thought about the **vulnerability** of IoT?

B: I don't know specifically, but I **deem** there must be a definite negative aspect of the IoT.

A: Right, and its vulnerability is necessary to understanding its darker aspect.

B: Is it related to data use?

A: You bet. Studies states that **devices** connected to the IoT can be **hacked**.

B: Really, how severe the problem is?

A: The numbers suggest that over 70% of all the devices in the IoT are vulnerable to cyber attacks.

B: That is awful, since data from many industries can be used to make huge danger.

A: Yes, one of the recent **scenarios** involves a healthcare **facility** where the clinical systems were targeted successfully.

B: Suchresults showed that a lot of data can be **breached** through the IoT networks.

A: Yes, that is why we have to attach real importance to such vulnerabilities so that IoT can benefit us in a safer way.

B: I totally agree with you.

 Key Words

vulnerability 弱点；脆弱

deem 认为

device 装置；设备

hack 非法侵入

scenario 情形；情景

facility 设备；设施

breach 违反；在…上打开缺口

Task 2 Act out the conversation with your partner based on the following clues.

A: _____①_____ ?

B: I don't know specifically, but I deem there must be a definite negative aspect of the IoT.

A: Right, and its vulnerability is necessary to understanding its darker aspect.

B: _____②_____ ?

A: Youbet. _____③_____ .

B: Really, how severe the problem is?

A: The numbers suggest that over 70% of all the devices in the IoT _____④_____ .

B: That is awful, since data from many industries can be used to make huge danger.

A: Yes, one of the recent scenarios involves a healthcare facility where the clinical systems were targeted successfully.

B: Such results showed that _____⑤_____ .

A: Yes, that is why we have to attach real importance to such vulnerabilities so that IoT can benefit us in a safer way.

B: I totally agree with you.

Conversation 3

A: Do you know what IoT means?

B: Is that an **acronym** for Internet of Things?

A: Right, IoT is the technology which is responsible to connect different devices and collect data.

B: What's the meaning and the purpose for this technology?

A: IoT technology is increasingly developed for **various** business purposes. Almost every **sector** is taking benefit of this trending technology in their business.

B: Really, how is that so, and what benefits are there?

A: First, it helps consumers achieve customer-centricity, which means the issue of customers can be **debugged** and their satisfaction levels can be improved.

B: So like in the scenario where mobile card readers are used, all transactions on smartphones can be **conducted** very smoothly.

A: Exactly. Meanwhile, by collecting rich data, organizations perform different analysis and do deep understanding so that they can improve their product quality.

B: And that means we as customers can receive better service.

A: Right, besides that, thanks to IoT, the increasing use of smart devices brings an **advancement** in smart device applications in different sectors.

B: That is how we are embracing more comfortable lives in all walks of life, including transportation, **hospitality**, healthcare, and education.

Key Words

acronym   首字母缩略词

various   各种各样的

sector   部门；领域

debug   调试

conduct   处理；实施

advancement   前进；进步

hospitality   酒店管理

Task 3   Act out the conversation with your partner based on the following clues.

A: Do you know what IoT means?

B: _____①_____ .

A: Right, IoT is the technology which is responsible to _____②_____ .

B: What's the meaning and the purpose for this technology?

A: _____③_____ for various business purposes. Almost every sector is taking benefit of this trending technology in their business.

B: Really, how is that so, and what benefits are there?

A: First, _____④_____ , which means the issue of customers can be debugged and their satisfaction levels can be improved.

B: So like in the scenario where mobile card readers are used, all transactions on smartphones can be conducted very smoothly.

A: Exactly. Meanwhile, by collecting rich data, organizations perform different analysis and do deep understanding so that they can improve their product quality.

B: And that means _____⑤_____ .

A: Right, besides that, thanks to IoT, the increasing use of smart devices brings an advancement in smart device applications in different sectors.

B: That is how we are embracing more comfortable lives in all walks of life,

including transportation, hospitality, healthcare, and education.

Part C   Passages

Listen to the following passages carefully and fill in the blanks with the information you've heard.

听力材料

Passage 1

On a basic level, IoT is used for collecting data about the world that would be very difficult or impossible for humans to without the aid of smart devices and _____ systems. derived from the data collected by these devices allow people to understand, monitor, and _____ to events or changes.

Passage 2

Businesses use IoT to _____ their supply _____, manage and improve customer experience, while smart _____ devices such as the Amazon Echo speaker, are now ubiquitous in homes due to the prevalence of low-cost and low-power _____ .

Passage 3

Beyond _____, IoT has helped make COVID disrupted chains more _____, automated activities in warehouses and on floors to help promote social distancing and provided safe access to industrial machines.

Passage 4

IBM has _____ allocated a _____ of around $200 million for their new global _____ in Munich. It will include _____ on IoT, Blockchain, artificial intelligence, and cybersecurity. Even Microsoft has _____ in the market to provide Platform as a Service (PaaS).

Passage 5

Implementing the IoT makes it easy to _____ physical devices from _____ locations simply by using a smart _____ . New business _____ in this field might seem daunting for many investors given the _____ response to a new product.

Section Ⅱ 　Reading

Passage 1

What is Internet of Things（IoT）?

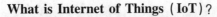

　　Simply put, the term Internet of Things **refers to** the entire network of physical devices, tools, **appliances**, equipment, **machinery**, and other smart objects that have the capability to collect data about the physical world and **transmit** that data through the Internet. ①

　　On a basic level, IoT is used for collecting data about the physical world that would be very difficult or impossible for humans to collect without the aid of smart devices and **monitoring** systems. ② **Insights derived** from the data collected by these devices allow people to understand, monitor, and react to events or changes. ③

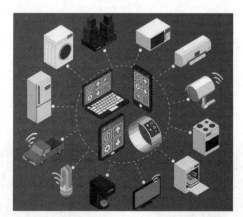

Internet of Things devices come in all different forms and are used for a **variety** of different things—from helping individuals with day-to-day tasks to helping large organizations **streamline** operations and meet business goals. ④ Things like smart appliances, smart lights, and smart door locks are all examples of IoT that you might find in someone's home. Examples of **commercial** and industrial IoT devices include things like smart factory or farming equipment, smart **vehicles**, and even entire connected factories, warehouses and buildings. ⑤

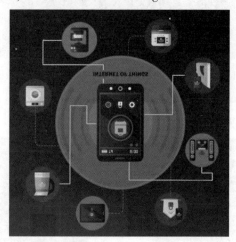

New Words and Expressions

refer to 涉及；指的是

appliance [ə'plaɪəns] *n.* 器具；应用；装置

machinery [mə'ʃiːnəri] *n.* 机械；机器

transmit [træns'mɪt] *v.* 传达；传送

monitor ['mɒnɪtə(r)] *v.* 监视；监督

insight ['ɪnsaɪt] *n.* 洞察力；见识；理解

derive [dɪ'raɪv] *v.* 获取；得自；起源

variety [və'raɪəti] *n.* 多样；种类

streamline ['striːmlaɪn] *v.* 使…合理化；使简化

commercial [kəˈmɜːrʃl]　*adj.* 商业的

vehicle [ˈviːɪkl]　*n.* 车辆

 Notes

1. Simply put, the term Internet of Things refers to the entire network of physical devices, tools, appliances, equipment, machinery, and other smart objects that have the capability to collect data about the physical world and transmit that data through the internet.

   Analysis：此句子中主句部分宾语比较长，而这些客体后又添加了用于修饰的定语从句。

   Translation：简单地说，物联网指的是由实体设备、工具、器具、设备、机械和其他智能物体组成的整个网络，这些物体能够收集有关物质世界的数据，并通过互联网传输这些数据。

2. On a basic level, IoT is used for collecting data about the physical world that would be very difficult or impossible for humans to collect without the aid of smart devices and monitoring systems.

   Analysis：本句中定语从句的先行词是一个整体 "data about the physical world"。核心词是 "data"。

   Translation：本质上说，物联网用于收集物理世界的数据，这些数据如果没有智能设备和监控系统的帮助，是很难或不可能被人类收集的。

3. Insights derived from the data collected by these devices allow people to understand, monitor, and react to events or changes.

   Analysis：本句中两次使用了过去分词做后置定语的用法，第一组被修饰的对象是 "insights"，第二组是 "data"。

   Translation：从这些设备收集的数据中得出的见解使人们能够理解、监控和应对事件或变动。

4. Internet of Things devices come in all different forms and are used for a variety of different things—from helping individuals with day-to-day tasks to helping large organizations streamline operations and meet business goals.

   Analysis：破折号起到解释说明作用，此处用来解释不同的事情都有哪些。

   Translation：物联网设备有各种不同的形式，有各种不同的用途，从帮助个人完成日常工作到帮助大型组织简化运营和实现业务目标。

5. Examples of commercial and industrial IoT devices include things like smart factory or farming equipment, smart vehicles, and even entire connected factories, warehouses and buildings.

   Translation：商业和工业物联网设备的例子包括智能工厂或农业设备、智能车辆，甚至全部互联的工厂、仓库和建筑物。

Exercises

Ⅰ. Answer the following questions according to the text.

1. What does the term Internet of Things refer to?

_____

_____

2. What capability do smart objects have?

_____

_____

3. What's IoT used for on a basic level?

_____

_____

4. What examples of IoT can you name that you might find in someone's home?

_____

_____

5. What do examples of commercial and industrial IoT devices include?

_____

_____

Ⅱ. Fill in the blanks with words according to the meaning of the article by memory.

Internet of Things _____ come in all different forms and are used for a _____ of different things—from helping _____ with day-to-day tasks to helping large organizations _____ operations and meet business goals. Things like smart _____, smart lights, and smart door locks are all examples of IoT that you might find in someone's home. Examples of commercial and _____ IoT devices include things like smart factory or farming equipment, smart _____, and even entire connected factories, warehouses and buildings.

Ⅲ. Fill in the blanks with the words given below. Change the forms when necessary.

insight	commercial	variety
transmit	appliance	refer to
derive	monitor	

1. Many smart objects have the capability to collect data about the physical world and _____ that data through the Internet.

2. Insights _____ from the data collected by these devices allow people to understand,

monitor, and react to events or changes.

3. IoT devices come in all different forms and are used for a _____ of different things.

4. Examples of _____ and industrial IoT devices include things like smart factory or farming equipment, smart vehicles.

5. Dish washer is one of the _____ .

6. IoT is used for collecting data about the physical world that would be very difficult or impossible for humans to collect without the aid of smart devices and _____ systems.

7. The book is filled with remarkable _____ .

8. The new law does not _____ land used for farming.

IV. Translate the following sentences into Chinese.

1. Insights derived from the data collected by these devices allow people to understand, monitor, and react to events or changes.

_____

_____

2. On a basic level, IoT is used for collecting data about the physical world that would be very difficult or impossible for humans to collect without the aid of smart devices and monitoring systems.

_____

_____

3. Internet of Things devices come in all different forms and are used for a variety of different things—from helping individuals with day-to-day tasks to helping large organizations streamline operations and meet business goals.

_____

_____

4. Things like smart appliances, smart lights, and smart door locks are all examples of IoT that you might find in someone's home.

_____

_____

5. Examples of commercial and industrial IoT devices include things like smart factory or farming equipment, smart vehicles, and even entire connected factories, warehouses and buildings.

_____

_____

_____

_____

Passage 2

Everyday Uses of IoT

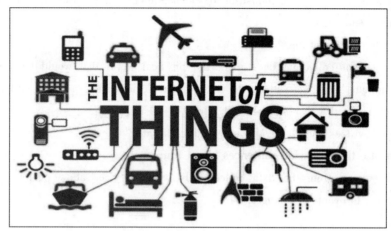

Connected devices fall into three domains: consumer IoT, such as **wearables**, **enterprise** IoT, which includes smart factories and **precision** agriculture, and public spaces IoT, such as waste management. ①

Businesses use IoT to **optimize** their supply chains, manage **inventory** and improve customer experience, while smart consumer devices such as the Amazon Echo speaker, are now **ubiquitous** in homes due to the **prevalence** of low-cost and low-power sensors. ②

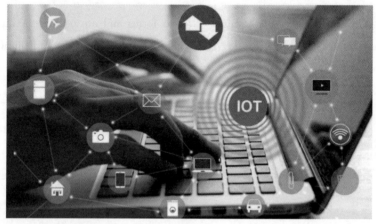

Cities have been **deploying** IoT technology for more than a decade-to streamline everything from **water meter** readings to traffic flow.

"In New York City, for example, every single building (so more than 817,000) was **retrofitted** with a wireless water meter, starting back in 2008, which replaced the manual system where you had to walk up to a meter read the numbers and generate bills that way," says Jeff Merritt, the World Economic Forum's head of IoT and Urban Transformation.

"Many cities now **leverage** license plate readers, traffic counters, red light cameras, radiation sensors and surveillance cameras to manage day-to-day operations. "③

In medicine, the IoT can help improve healthcare through real-time remote patient monitoring, robotic surgery and devices such as smart inhalers.

In the past 12 months, the role of the IoT in the COVID-19 **pandemic** has been invaluable.

"IoT applications such as connected thermal cameras, contact tracing devices and health-monitoring wearables are providing critical data needed to help fight the disease, while temperature sensors and parcel tracking will help ensure that sensitive COVID-19 vaccines are distributed safely, "④ according to the Forum's State of the Connected World report.

Beyond healthcare, IoT has helped make COVID-disrupted supply chains more **resilient**, automated activities in warehouses and on factory floors to help **promote** social distancing and provided safe remote access to industrial machines. ⑤

 New Words and Expressions

wearable ['weərəbl]　n. 衣服，穿戴设备

enterprise ['entəpraɪz] *n.* 企业；事业

precision [prɪ'sɪʒn] *n.* 精确；精密度

optimize ['ɒptɪmaɪz] *v.* 使完善；使优化

inventory ['ɪnvəntri] *n.* 详细目录；存货

ubiquitous [juː'bɪkwɪtəs] *adj.* 到处存在的

prevalence ['prevələns] *n.* 传播；流行；普及

deploy [dɪ'plɔɪ] *v.* 部署；配置

water meter 水表（量水器）

retrofit ['retrəʊfɪt] *v.* 式样翻新；改进

leverage ['levərɪdʒ] *v.* 利用，借助；以…为杠杆

pandemic [pæn'demɪk] *n.* 流行病

resilient [rɪ'zɪliənt] *adj.* 适应力强的；有弹力的

promote [prə'məʊt] *v.* 促进；提升

Notes

1. Connected devices fall into three domains: consumer IoT, such as wearables, enterprise IoT, which includes smart factories and precision agriculture, and public spaces IoT, such as waste management.

   Translation：互联设备分为三个领域：消费物联网，如可穿戴设备；企业物联网，包括智能工厂和精准农业；公共空间物联网，例如废物管理。

2. Businesses use IoT to optimize their supply chains, manage inventory and improve customer experience, while smart consumer devices such as the Amazon Echo speaker, are now ubiquitous in homes due to the prevalence of low-cost and low-power sensors.

   Analysis："while" 引导让步状语从句，表示"而…"。

   Translation：企业使用物联网来优化供应链、管理库存和改善客户体验，而智能消费设备，像亚马逊"回声"语音器，由于低成本和低功耗传感器的普及，现在在家庭中无处不在。

3. Many cities now leverage license plate readers, traffic counters, red light cameras, radiation sensors and surveillance cameras to manage day-to-day operations.

   Translation：许多城市现在利用车牌阅读器、交通计数器、红灯摄像机、辐射传感器和监控摄像机来管理日常运营。

4. IoT applications such as connected thermal cameras, contact tracing devices and health-monitoring wearables are providing critical data needed to help fight the disease, while temperature sensors and parcel tracking will help ensure that sensitive COVID-19 vaccines are distributed safely.

   Translation：联网热成像相机、接触追踪设备和健康监测可穿戴设备等物联网应用正在提供帮助抗击疾病所需的关键数据，而温度传感器和包裹跟踪将有助于确保敏感

的 COVID-19 疫苗得到安全分发。

5. Beyond healthcare, IoT has helped make COVID-disrupted supply chains more resilient, automated activities in warehouses and on factory floors to help promote social distancing and provided safe remote access to industrial machines.

Analysis：本句中三组谓语动词并列，最后一组用"and"连接。

Translation：除了医疗保健，物联网还帮助提高了受疫情影响的供应链的抵御能力，自动化了仓库和工厂车间的活动，从而有效促进了社交距离实现，并为工业机器提供了安全的远程访问。

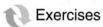 Exercises

Answer the following questions according to the text.

1. What are the three domains that connected devices fall into?

_____

_____

2. Whatdo businesses use IoT to do?

_____

_____

3. How long have cities been deploying IoT technology?

_____

_____

4. What can IoT do in medicine?

_____

_____

5. What IoT applications are providing critical data needed to help fight the COVID-19?

_____

_____

_____

_____

Passage 3

Interesting Facts about the IoT

The Internet of Things is still in its growth stage. It is still an unfamiliar term in many parts of the world. But there are already Internet of Things companies out there you might want to know and here is a list of facts which you should definitely keep in mind.

1. Widespread adoption by Corporate Companies

Many big-time corporates and well known multinational companies have turned to increasingly adopt the Internet of Things in the regular business. [1] This is highly evident from the financial statistics and research **expenses** in this field.

IBM has reportedly **allocated** a **budget** of around $200 million for their new global headquarters in Munich. It will include projects on IoT, Block Chain, Artificial Intelligence, and cybersecurity. Even Microsoft has entered in the market to provide Platform as a Service (PaaS).

As reported by the Business Insider, the financial **institution**, Morgan Stanley predicts the devices connected to the Internet of Things will cross the 75 billion **thresholds** by the year 2020. [2]

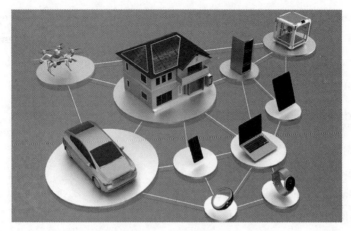

2. IoT has been around for the last 3 to 4 decades

The IoT trend is picking up very fast in the last couple of years. However, the IoT has been around since the early 1980's. As reported, the first instance of a physical device linked to the IoT was in the year 1982.

A modified coke machine was connected to the IoT to report the status of inventory in the machine. In the early phases, the IoT had restricted scope because of the poorly developed technology of that time.

But now, owing to the fast-paced internet access, availability of smartphones and mobile technology, the Internet of Things will increasingly **manifold**. ③

3. IoT is a promising business opportunity

IoT **ventures** are extremely **alluring** and offer very **lucrative** returns to investors. The Internet of Things has the ability to reshape everyday life.

Implementing the IoT makes it easy to operate physical devices from remote locations simply by using a smart device. New business opportunities in this field might seem **daunting** for many investors given the uncertain response to a new product. ④

But as per market statistics, almost 94% of the business units that have adopted the IoT have seen a **substantial** return on investments. General Electric, one of the **prominent** business house, has stepped into the IoT space for building a jet engine technology. GE has allocated a budget of over $60 trillion investments in the next 15 years.

New Words and Expressions

adoption [əˈldɒpʃn]   *n.* 采用；采纳

expense [ɪkˈlspens]   *n.* 费用；花费

allocate [ˈlæləkeɪt]   *v.* 分派；分配

budget [ˈlbʌdʒɪt]   *n.* 预算

institution [ˌɪnstɪˈltuːʃn]   *n.* 机构；制度

threshold [ˈlθreʃhoʊld]   *n.* 门槛；界限

manifold [ˈlmænɪfoʊld]   *v.* 繁殖；增多

venture [ˈlventʃər]   *n.* 冒险（事业）

allure [əˈllʊr]   *v.* 诱惑；吸引

lucrative [ˈlluːkrətɪv]   *adj.* 合算的；获利的

daunting [ˈdɔːntɪŋ]   *adj.* 令人畏惧的

substantial [səbˈlstænʃl]   *adj.* 大量的；可观的

prominent [ˈlprɒmɪnənt]   *adj.* 杰出的；突出的

Notes

1. Many big-time corporates andwell known multinational companies have turned to increasingly adopt the Internet of Things in the regular business.

   Translation：许多大型企业和知名跨国公司已转向在常规业务中越来越多地采用物联网。

2. As reported by the *Business Insider*, the financial institution, Morgan Stanley predicts the devices connected to the Internet of Things will cross the 75 billion thresholds by the year 2020.

   Analysis："As reported by the Business Insider" 作状语，此处是 "as" 引导的方式状语从句，主语和系动词一并省略，相当于 "As it is reported by the Business Insider"。

   Translation：据金融机构《商业内幕》报道，摩根士丹利预测，到 2020 年，连接物联网的设备将超过 750 亿台。

3. But now, owing to the fast-paced internet access, availability of smartphones and mobile technology, the Internet of Things will increasingly manifold.

   Analysis："owing to…" 表示 "因为，由于"，作原因状语。

   Translation：但现在，由于快速互联网的接入、智能手机和移动技术的普及，物联网将越来越多样化。

4. New business opportunities in this field might seem daunting for many investors given the uncertain response to a new product.

Analysis："given…" 表示"鉴于…"，作为原因状语放在句尾。

Translation：考虑到对新产品的不确定反应，这一领域的新商机对许多投资者来说可能令人望而生畏。

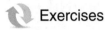 Exercises

Translate the following short passages into Chinese.

1. As reported by the Business Insider, the financial institution, Morgan Stanley predicts the devices connected to the Internet of Things will cross the 75 billion thresholds by the year 2020.

_____

_____

_____

2. A modified coke machine was connected to the IoT to report the status of inventory in the machine. In the early phases, the IoT had restricted scope because of the poorly developed technology of that time.

_____

_____

_____

3. Implementing the IoT makes it easy to operate physical devices from remote locations simply by using a smart device. New business opportunities in this field might seem daunting for many investors given the uncertain response to a new product.

_____

_____

_____

_____

Section Ⅲ　Grammar

Participle（分词）

一、分词的基本概念

分词也是非谓语动词之一，是英语中一种常见的语法形式。分词通常由动词的-ing 或-ed 形式构成，分别称为现在分词和过去分词：

1. 现在分词：动词原形 + ing（同动名词形式）

非谓语动词形式	时态/语态	主动形式	被动形式
现在分词	一般式	doing	being done
	完成式	having done	having been
	完成进行式	having been doing	（基本不用）
否定形式：not + 现在分词			

2. 过去分词：（规则动词）动词原形 + ed；（不规则动词）动词过去分词

二、分词在句中所能充当的成分

	主语	宾语	表语	定语	状语	补语
现在分词	×	×	√	√	√	√
过去分词	×	×	√	√	√	√

三、分词的用法及其逻辑主语

分词的基本用法：现在分词表主动和进行，过去分词表被动和完成。

句中的作用	逻辑主语	分词与逻辑主语的关系	举　　例
作定语（前置或后置定语）	被修饰的名词	主动关系：用现在分词	an interesting story a girl standing under the tree
		被动关系：用过去分词	a broken cup a book written by him
作宾语补足语	宾语	主动关系：用现在分词	I heard someone knocking at the door. I always found him reading in the morning.
		被动关系：用过去分词	I found everything put in good order. I once heard this song sung in English.
作状语	全句的主语	主动关系：用现在分词	Being very busy, my father did not come back yesterday.
		被动关系：用过去分词	Helped by the teacher, I have made much progress in English study.

四、现在分词、过去分词和不定式作宾语补足语的区别

	形式	语态	内容	被动语态
现在分词	v +ing	与宾语是主动关系	动作正在进行还没有结束	没有变化
过去分词	v + ed 或不规则变化	与宾语是被动关系	动作已经结束	没有变化
不定式	不带 to，动词原形	与宾语是主动关系	动作发生了，全过程已结束	to 要还原

五、独立主格结构

这里提到的分词用法通常会有一个逻辑主语，但有时分词也可以与自己的独立主语构成独立主格结构。这种独立主语通常是名词或代词，并且出现在分词前面。

独立主格结构通常只作为句子的状语使用。分词的形式通常取决于它们表示的动作与句子的谓语动词之间的时间关系。对于独立主格中使用现在分词或过去分词的问题，应根据它们的主语和表示的动作是主动还是被动来确定。

Eg. The bell ringing, we all stopped talking.

注：

①独立结构中的 being 或 having been 常可省去，如：The meeting (being) over, all left the room.

②作伴随状语的独立结构常可用 with 短语来代替，如：

Her arms folded, she leaned against the wall.

With her arms folded, she leaned against the wall.

 Exercises

Choose the best answer.

1. You need to consider either a paid or volunteer bridge job while _____ full-time work.

    A. to seek   B. sought   C. seek   D. seeking

2. _____ the investigation, the committee published the report on the cause of the accident.

    A. Completed   B. Completing

    C. Having completed   D. To have completed

3. It seems that the suggestion _____ at yesterday's meeting doesn't sound practical.

    A. make   B. made   C. to be made   D. making

4. The sales manager resigned last Monday after _____ to take a pay cut by the company.

    A. be asked   B. being asked   C. was asked   D. is asked

5. _____ by all the team numbers, they finally got the big project for their company.

    A. To be supported   B. Having supported

    C. Supporting   D. Supported

6. This dictionary is a three-common word collection with samples of _____ and spoken

language.

    A. writing   B. write   C. wrote   D. written

7. All our spare parts are guaranteed if you have your car _____ with us each year.

    A. serviced   B. be servicing   C. to service   D. servicing

8. The course comes in three books of case studies, _____ a variety of business activities in different parts of the world.

    A. being covered   B. covering   C. cover   D. covered

9. _____ great losses in the financial crises, the company closed down last year.

    A. Being suffered   B. To suffer   C. Having suffered   D. Suffered

10. _____ with the size of the whole earth, the biggest ocean does not seem big at all.

    A. Compare   B. When comparing

    C. Comparing   D. When compared

Section IV   Writing

A Tour Commentary (导游词)

导游词是指导游为游客在旅游途中提供的解说词，是导游员同游客交流思想，向游客传播文化知识的工具，也是应用写作研究的文体之一。导游词通常包括景点的历史、文化、艺术、建筑、地理等方面的知识，帮助游客更好地理解和欣赏所参观的景点。

Part 1   Sample

The following is a tour commentary. Please read and try to understand it.

MountQianshan Tour Commentary

Ladies and gentlemen,

Good morning. Today we are going to visit one of the famous scenic spots in Liaoning Province, Mount Qianshan.

Located in the middle of Liaoning Province, Qianshan Mountain Scenic Spot is 17 kilometers away in the east of Anshan City, and 30 kilometers away in the south of Liaoyang City. The total area of the spot is 125 square kilometers, and there are still 72 square kilometers under construction. Mt. Qianshan means a mountain with a thousand peaks, and actually Qianshan has 999. It is well-known for the marvelous peaks, the precipitous rocks and quaint temples, and enjoys the fame of "The Northeast Pearl".

Qianshan Mountain was also called Thousand Lotus Mountain. She had age-old history. It was said there were some temple buildings in the Shui and Tang dynasty about 1300 years ago. After Ming and Qing dynasty it gradually became the center of Taoism and Buddhism.

OK, ladies and gentlemen, so much for Qianshan. Thank you for your listening. I'm looking forward to your next visit.

Part 2   Template

从上面的样例可以看出英语导游词的书写结构为：

第一部分，向游客发出问候语。简短的问好之后，导游可以简要介绍旅游行程的背景和重点，引起游客的兴趣和注意。

第二部分，也是最重要的部分，是景点的详细介绍。导游会介绍当前所在的景点，包括景点的历史、文化、建筑、艺术等方面的知识，以及景点的特点和特色。

除此之外，在此部分可以添加介绍景点的亮点，包括值得关注的历史事件、文化遗产、建筑特色等，以便游客更深入地了解景点的精华部分。

最后一部分是小结，导游会在讲解结束时进行总结和概括，再次强调本次旅游的重点和亮点，并提供必要的建议和提示，以便游客更好地享受旅游体验。

Ladies and gentlemen,

Welcome to _____ . I am very delighted to be your guide here. Please allow me to give you a brief introduction of _____ .

_____ in the east of _____ Province, is famous for its _____ . _____ is a wonderful city.

Every summer holiday, thousands of people from all over the world come to visit the city. You can walk along the beaches, go swimming in the sea, or do some shopping in the stores.

Great changes have taken place in _____ in the recent years. As an international port city, it has played an important role in the development of the foreign trade of our city.

Thanks for listening. Wish you a good time in _____!

Part 3 Useful Patterns

1. Please follow me.	1. 请跟我走。
2. We are about to pass through…	2. 我们马上就要走过…
3. We have now reached the summit of…	3. 我们现在到了…的顶部
4. Now, I will give you a brief introduction to the history of…	4. 现在我给大家简单介绍一下…的历史。
5. To the north of…lies…	5. …在…的北面，是…
6. Looking back, we can see…	6. 往后看，我们就可以看到…
7. The history of this place can be traced back to…years ago.	7. 这地方的历史可以追溯到…年前。
8. Once we climb to the top of…, we can see…	8. 爬到…的顶部，我们就能看到…
9. Ladies and gentlemen, please pay attention!	9. 各位，请注意！

Part 4 Exercises

I. Translate the following tour commentary into Chinese.

Ladies and gentlemen,

I'm pleased to serve as your guide today. What is the most representative place in Beijing? The answers are various. But Tian, anmen Square is unarguably on the top list.

Tian'anmen is located in the center of Beijing. It was first built in 1417 and named Chengtianmen. At the end of the Ming Dynasty, it was seriously damaged by war. When it was rebuilt under the Qing in 1651, it was renamed Tian'anmen, and served as the main entrance to the Imperial City, the administrative and residential quarters for court officials and retainers. The southern sections of the Imperial City wall still stand on both sides of the Gate. The tower at the top of the gate is nine-room wide and five-room deep. According to the Book of Changes, the two numbers nine and five, when combined, symbolize the supreme status of a sovereign. During the Ming and Qing dynasties, Tian' anmen was the place where state ceremonies took place. This was the place when in 1949, from a rostrum on Tian'anmen, Chairman Mao announced the establishment of the People's Republic of China. Tiananmen Square is circled by Tiananmen on its north; the Great Hall of the People on its west; on the east of Tian'anmen Square lies the National Museum of China; there are Monument to the People's Heroes and Chairman Mao's Mausoleum on the south. At sunrise and sunset the raising and lowering ceremony of the Chinese National Flag is well worth seeing. The young troops perform very well. Make sure to be there 30 minutes earlier to get a good standing point.

This is the introduction of Tian'anmen Square. I hope it can help you to have a better understanding of this place. Thank you.

II. Write a tour commentary according to the information given in Chinese.

假如你是一名导游，陪同某外国旅游团游览长城。车抵长城，下车前你准备对大家作必要的介绍和交代。请拟一份讲话稿。讲话稿必须包括以下内容：

1. 简单介绍长城（世界上最长的城墙，有两千多年的历史，是世界奇观之一。）

2. 在长城逗留两个半小时，11 点离开。

3. 游览车在入口处等候，记住车号，准时返回。

4. 贵重物品随身携带，下车前关上车窗。

5. 游览时注意安全，祝大家玩得愉快。

注意：不要逐句翻译，字数 100 个单词左右。

Unit 9

Block Chain

Unit Goals

After learning this unit, you will be able to:

- talk about block chain
- understand the passages and grasp the key words and expressions
- review the grammar of the nominal clause
- write an e-mail

Section Ⅰ　Listening & Speaking

fundamental　基本的；根本的	block chain　区块链
ledger　总账；账目	immutable　不可变的；不变的
facilitate　促进；帮助；使…容易	transaction　交易；办理
asset　资产	intangible　无形的
track　追踪；跟踪	remarkable　非凡的；精彩的
distributed　分布的；分散式的	eliminate　除去；剔除
tamper　干预；篡改	visible　看得见的；可见的
execute　执行；实行	sequence　顺序
insert　插入；嵌入	verification　确认；查证
hence　因此；从此以后	deliver　递送；交付；带来
malicious　恶意的；恶毒的	

Part B　Conversations

Read the conversations carefully and then complete the communicative tasks.

Conversation 1

A：Do you know any **fundamental** information about **block chain**?

B：I only know that it is defined as a secure **ledger**.

A：Yep，block chain is a shared，**immutable** ledger that **facilitates** the process of recording **transactions** and tracking assets in a business network.

B：You mean tangible **assets** like a house, car, cash, land?

A：Not only those, but also **intangible** assets such as intellectual property, patents, copyrights, branding.

B：Oh, that means virtually anything of value can be tracked and traded on a block chain network.

A：Yes, that's why it really helps reducing risk and cutting costs for all involved.

B：I didn't imagine such impressive advantages of it.

A：Even better, block chain is ideal for delivering information that all business runs on, which relies on receiving speed and accuracy.

B：Why is that so?

A：Because it provides immediate, shared and completely transparent information stored on an immutable ledger.

B：What if someone else intends to steal the stored information?

A：Don't worry. Such ledgers can be accessed only by permitted network members.

Key Words

fundamental 基本的；根本的
block chain 区块链
ledger 总账；账目
immutable 不可变的；不变的
facilitate 促进；帮助；使…容易
transaction 交易；办理
asset 资产
intangible 无形的

Task 1　Act out the conversation with your partner based on the following clues.

A：Do you know any fundamental information about block chain?

B：_____①_____.

A：Yep, block chain is a shared, immutable ledger that facilitates the process of recording transactions and _____②_____.

B：You mean tangible assets like a house, car, cash, land?

A：Not only those, but also intangible assets such as _____③_____.

B：Oh, that means virtually anything of value can be tracked and traded on a block chain network.

A：Yes, that's why it really helps reducing risk and cutting costs for all involved.

B：_____④_____.

A: Even better, block chain is ideal for delivering information that all business runs on, which relies on receiving speed and accuracy.

B: _____⑤_____?

A: Because it provides immediate, shared and completely transparent information stored on an immutable ledger.

B: What if someone else intends to steal the stored information?

A: Don't worry. Such ledgers can be accessed only by permitted network members.

Conversation 2

A: It's said that block chain network can **track** orders, payments, accounts, production and much more.

B: Besides that fact, since members share a single view of the truth, they can see all details of a transaction end to end.

A: That's **remarkable**. I wonder what elements support such brilliant functions.

B: There are mainly three key elements of a block chain.

A: Please explain to me what they are.

B: The first one is **distributed** ledger technology. All network participants have access to the distributed ledger and its immutable record of transactions.

A: I know. With this shared ledger, transactions are recorded only once, **eliminating** the duplication of effort.

B: Right, along with that it is immutable records, which is the second key element.

A: That means no participant can change or **tamper** with a transaction after it's been recorded to the shared ledger, right?

B: Bingo. If a transaction record includes an error, a new transaction must be added to reverse the error, and both transactions are then **visible**.

A: The third key element is smart contracts. To speed transactions, a set of rules — called a smart contract — is stored on the block chain and **executed** automatically.

B: That is out of my knowledge zone.

 Key Words

track   追踪；跟踪

remarkable   非凡的；值得注意的

distributed   分布的；分散式的

eliminate   除去；剔除

tamper 干预；篡改
visible 看得见的；可见的
execute 执行；实行

Task 2 Act out the conversation with your partner based on the following clues.

A： It's said that block chain network can track orders, payments, accounts, production and much more.

B： Besides that fact, ____①____, they can see all details of a transaction end to end.

A： That's remarkable. ____②____.

B： There are mainly three key elements of a block chain.

A： Please explain to me what they are.

B： ____③____. All network participants have access to the distributed ledger and its immutable record of transactions.

A： I know. With this shared ledger, ____④____, eliminating the duplication of effort.

B： Right, along with that is immutable records, which is the second key element.

A： That means no participant can change or tamper with a transaction after it's been recorded to the shared ledger, right?

B： Bingo. If a transaction record includes an error, a new transaction must be added to reverse the error, ____⑤____.

A： The third key element is smart contracts. To speed transactions, a set of rules — called a smart contract — is stored on the block chain and executed automatically.

B： That is out of my knowledge zone.

Conversation 3

A： Can you explain to me how block chain works?

B： As each transaction occurs, it is recorded as a "block" of data.

A： You mean the data block can record every movement of your asset.

B： Yep, the data block can record the information of your choice: who, what, when, where, how much and even the condition.

A： That's amazing!

B： Moreover, these blocks form a chain of data as an asset moves from place to place or ownership changes hands.

A： Which is to say, each block is connected to the ones before and after it?

B： That's right. The blocks confirm the exact time and **sequence** of transactions, and the blocks link securely together.

A： That's how it prevents any block from being altered or a block being **inserted** between two existing blocks.

B：Exactly. Meanwhile, each additional block strengthens the **verification** of the previous block and **hence** the entire block chain.

A：So, this renders the block chain tamper-evident, **delivering** the key strength of immutability.

B：Right, this removes the possibility of tampering by a **malicious** actor, and builds a ledger of transactions you and other network members can trust.

Key Words

sequence　顺序

insert　插入；嵌入

verification　确认；查证

hence　因此；从此以后

deliver　递送；交付；带来

malicious　恶意的；恶毒的

Task 3　Act out the conversation with your partner based on the following clues.

A：_____①_____ ?

B：As each transaction occurs, _____②_____ .

A：You mean the data block can record every movement of your asset.

B：Yep, the data block can record the information of your choice：who, what, when, where, how much and even the condition.

A：_____③_____ !

B：Moreover, these blocks form a chain of data as an asset moves from place to place or ownership changes hands.

A：Which is to say, _____④_____ ?

B：That's right. The blocks confirm the exact time and sequence of transactions, and the blocks link securely together.

A：That's how it prevents any block from being altered or a block being inserted between two existing blocks.

B：Exactly. Meanwhile, each additional block strengthens the verification of the previous block and hence the entire block chain.

A：So, this renders the block chain tamper-evident, _____⑤_____ .

B：Right, this removes the possibility of tampering by a malicious actor, and builds a ledger of transactions you and other network members can trust.

听力材料

Part C　Passages

Listen to the following passages carefully and fill in the blanks with the information you've heard.

Passage 1

Every computer software that uses ablock chain, will give its users a _____ and a private key. These are again just like _____ ; they are a _____ sequence of alphabets and numbers that are _____ by the software itself. Every user has to keep their private key securely and not _____ it to anyone. The public key, on the other hand, can be revealed to everyone.

Passage 2

A hash is a unique _____ of letters and numbers. It is like a _____ for the data in a _____ and it is always unique to every block in the block chain. When the _____ in a block changes, the _____ will also change.

Passage 3

_____ the example of a mailbox. The _____ key is like your mailbox which everyone knows about, and can _____ you messages. The _____ key, on the other hand, is like the key to that mailbox. Only you _____ it, and only you can read the messages inside.

Passage 4

The technology can work for almost every type of involving value, including money, _____ and property. Its uses are almost limitless: from _____ taxes to enabling to send money back to family in countries where banking is difficult.

Passage 5

_____ only a very small _____ of _____ GDP (around 0.025%, or \$20 billion) is held in the _____, according to a _____ by the World Economic Forum's Global Agenda Council.

Section II   Reading

Passage 1

Block Chain

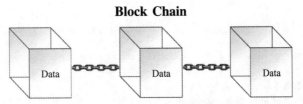

Ablock chain is a method of storing data. Data is stored in blocks which are linked to the previous block.

But what does a "Block" look like?

Each "block" **contains**

① Data of transactions

② A **unique** fingerprint for all the data in the block called a **hash**

③ A hash of the previous block's data

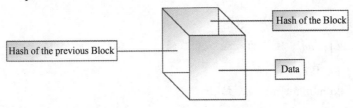

What does each of these items mean?

Data in the block usually consists of transactions. A block can contain hundreds of transactions. Alice sending Bob $100 is an example of a transaction in a block.

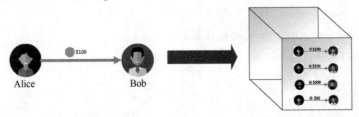

A hash is a unique **combination** of letters and numbers. It is like a **fingerprint** for the data in a block and it is always unique to every block in the block chain. ① When the data in a block changes, the hash will also change.

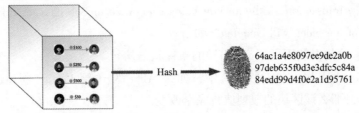

Hence in a transaction, if the amount being sent Alice to Bob changes from $50 to $100, the hash of the block will completely change. ②

A block also contains the hash of the **previous** block. Hence forming a chain structure. Combining the above three together, this is what a block chain will look like:

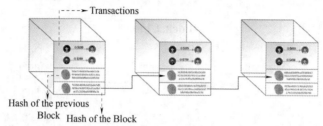

Now if a transaction in any block changes, the hash of the block will change. When the hash of the block changes, the next block will show a **mismatch** with the previous hash that was recorded by it. ③

This givesblock chain the **property** of being tamper-**resistant** as it becomes very easy to **identify** when data in a Block has changed. ④

 ## New Words and Expressions

contain [kən'teɪn] *v.* 容纳；包含

unique [ju'niːk] *adj.* 独特的；独一无二的

hash [hæʃ] *n.* 散列

combination [ˌkɒmbɪ'neɪʃn] *n.* 结合

fingerprint ['fɪŋgəprɪnt] *n.* 指纹；特征

previous ['priːviəs] *adj.* 以前的

mismatch ['mɪsmætʃ] *n.* 错配；不匹配

property ['prɒpəti] *n.* 财产；所有物

resistant [rɪ'zɪstənt] *adj.* 抵抗的；反抗的

identify [aɪ'dentɪfaɪ] *v.* 识别，认出

Notes

1. It is like a fingerprint for the data in a block and it is always unique to every block in theblock chain.

   Translation：它就像区块中数据的指纹，对于区块链中的每个区块都是唯一的。

2. Hence in a transaction, if the amount being sent Alice to Bob changes from $50 to $100, the hash of the block will completely change.

   Analysis：此句中使用了"if"引导的条件状语从句。

   Translation：因此，在一个交易中，如果 Alice 发送给 Bob 的金额从 50 美元变为 100 美元，那么该区块的散列值将完全改变。

3. When the hash of the block changes, the next block will show a mismatch with the previous hash that was recorded by it.

   Analysis：句尾处 "that was recorded by it" 中的 "it" 代指下一个区块。

   Translation：当区块的散列值发生变化时，下一个区块将显示与该区块链所记录的前一个散列值的不匹配。

4. This gives block chain the property of being tamper-resistant as it becomes very easy to identify when data in a Block has changed.

   Analysis："property" 在这里是名词，表示 "特性，性质" 的意思。

   Translation：这使区块链具有防篡改的特性，因为当区块中的数据发生变化时，它很容易识别出来。

Exercises

Ⅰ. Answer the following questions according to the text.

1. Sincea block chain is a method of storing data, where is data stored?

　_____

　_____

2. What does each block contain?

　_____

　_____

3. What does data in the block usually consist of ?

　_____

　_____

4. What's a hash?

　_____

　_____

5. What will happen if a transaction in any block changes?

　_____

　_____

Ⅱ. Fill in the blanks with words according to the meaning of the article by memory.

Now if a _____ in any block changes, the _____ of the block will change. When the hash of the block changes, the _____ will show a _____ with the hash that was recorded by it.

This givesblock chain the _____ of being _____ as it becomes very easy to when data in a Block has changed.

III. Fill in the blanks with the words given below. Change the forms when necessary.

combination	previous	unique
property	contain	identify
fingerprint	mismatch	

1. This gives block chain the _____ of being tamper-resistant as it becomes very easy to identify when data in a Block has changed.

2. It is like a _____ for the data in a block and it is always unique to every block in the block chain.

3. She _____ her son among a lot of children.

4. This book _____ all the information you need.

5. When the hash of the block changes, the next block will show a _____ with the previous hash that was recorded by it.

6. A _____ fingerprint for all the data in the block called a hash.

7. A hash is a unique _____ of letters and numbers.

8. A block also contains the hash of the _____ block.

IV. Translate the following sentences into Chinese.

1. Data is stored in blocks which are linked to the previous block.

_____

_____

2. A hash is a unique combination of letters and numbers.

_____

_____

3. It is like a fingerprint for the data in a block and it is always unique to every block in the block chain.

_____

_____

4. When the hash of the block changes, the next block will show a mismatch with the previous hash that was recorded by it.

_____

_____

5. This givesblock chain the property of being tamper-resistant as it becomes very easy to

identify when data in a Block has changed.

Passage 2

How to Create a Transaction in a Block Chain?

Every computer software that uses a block chain, will give its users a public key and a **private** key. These are again just like hashes; they are a **random** sequence of **alphabets** and numbers that are generated by the software itself. Every user has to keep their private key **securely** and not **reveal** it to anyone. The public key, on the other hand, can be revealed to everyone.

PRIVATE KEY
6831728990636725551934513790552817929570764
757855868444051228709791946 7220420

PUBLIC KEY
044e554e13e016a83a958197cf3b8622b9afc5b9ea04
bdf37elef20a2dabcfa7d180ba760ec74408abadd246
8bc5415d67305dd679d4bd1610c72fOaff57dc1ab3

Consider the example of a mailbox. The public key is like your mailbox which everyone knows about, and can drop you messages. The private key, on the other hand, is like the key to that mailbox. Only you own it, and only you can read the messages inside.

Both public and private keys have a unique property. The private key can be used to sign any message to create a **digital** signature. A digital signature is yet another **sequence** of characters and numbers. But there's a catch!

All digital signatures can be **verified** using the **corresponding** public key. [1] This means anyone who has a digital signature can verify whether a person truly signed the message, using the signer's public key. [2]

Both these keys, combined with message signing to create digital signatures can be called the **cryptography** in block chains.

Too complex?

Let's break it down step-by-step with an example.

Alice wishes to record the message that she sent $100 to Bob on ablock chain.

1. She writes the message and signs it using her private key to create a digital signature. Her message combined with the signature is a transaction.

2. The software Alice uses **broadcast** her transaction to everyone in the peer-to-peer network. ③

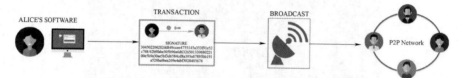

3. Everyone in the P2P network first verifies her transaction signature, to see if Alice is the one who really signed that message. They do so using Alice's public key which everyone knows.

4. Once verified, the P2P network includes Alice's transaction on a block in a block chain. ④

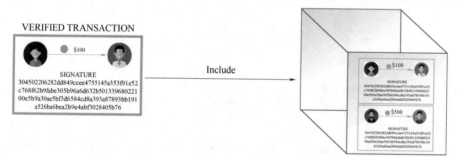

5. When the P2P network reaches consensus, after verifying all transactions, the block with Alice's transaction gets included in the block chain!⑤

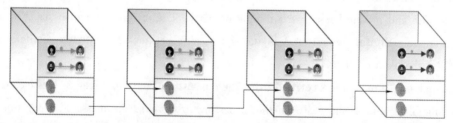

Once included, Alice's transaction cannot be changed by anyone so easily!

New Words and Expressions

private [ˈpraɪvət]　*adj.* 私人的；个人的

random ['rændəm]　*adj.* 任意的；随机的

alphabet ['ælfəbet]　*n.* 字母

securely [sɪ'kjʊəlɪ]　*adv.* 安全地；牢固地

reveal [rɪ'viːl]　*v.* 显示；透露；揭示

digital ['dɪdʒɪtl]　*adj.* 数字的；数码的

sequence ['siːkwəns]　*n.* 顺序；一系列

verify ['verɪfaɪ]　*v.* 核实；证明

corresponding [ˌkɒrə'spɒndɪŋ]　*adj.* 相当的；对应的

cryptography [krɪp'tɒgrəfi]　*n.* 密码系统

broadcast ['brɔːdkɑːst]　*v.* 广播；散布

peer-to-peer　同辈对同辈；对等网络

Notes

1. All digital signatures can be verified using the corresponding public key.

   Translation：所有数字签名都可以使用相应的公钥进行验证。

2. This means anyone who has a digital signature can verify whether a person truly signed the message, using the signer's public key.

   Analysis：本句中复合使用了定语从句，宾语从句和现在分词作状语表示伴随动作。

   Translation：这意味着任何拥有数字签名的任何人都可以使用签名者的公钥来验证某人是否真正签署了消息。

3. The software Alice uses broadcast her transaction to everyone in the peer-to-peer network.

   Analysis："Alice uses" 部分为定语从句，因为先行词在该从句中作宾语，所以连接词 "that" 在此处省略了。

   Translation：Alice 使用的软件向对等网络中的每个人广播她的交易。

4. Once verified, the P2P network includes Alice's transaction on a block in a block chain.

   Analysis："Once verified" 在这里作状语，也可以理解成状语从句 "Once it is verified" 把主语和系动词一起省略。

   Translation：一旦验证，P2P 网络就会将 Alice 的交易纳入区块链中的一个区块上。

5. When the P2P network reaches consensus, after verifying all transactions, the block with Alice's transaction gets included in the block chain!

   Analysis："When the P2P network reaches consensus" 和 "after verifying all transactions" 在这里均作时间状语，前者是时间状语从句，后者是介词短语。

   Translation：当 P2P 网络达成共识时，在验证所有交易后，带有 Alice 交易的区块将被纳入区块链！

 **Exercises**

Answer the following questions according to the text.

1. What will every computer software that uses a block chain give its users?

_____

_____

2. Can the public key be revealed to everyone?

_____

_____

3. Whats a digital signature?

_____

_____

4. How candigital signatures be verified?

_____

_____

5. How can people the P2P network verifie Alice's transaction signature?

_____

_____

Passage 3

Details about Block Chain

Many people know it as the technology behind Bitcoin, but block chain's potential uses **extend** far beyond digital **currencies**. ①

Its admirers include Bill Gates and Richard Branson, and banks and insurers are falling over one another to be the first to work out how to use it.

So what exactly is block chain, and why are Wall Street and **Silicon Valley** so excited about it?

What is block chain?

Currently, most people use a trusted middleman such as a bank to make a transaction. But block chain allows consumers and suppliers to connect directly, **removing** the need for a third party. [2]

Using cryptography to keep **exchanges** secure, block chain provides a **decentralized** database, or "digital ledger", of transactions that everyone on the network can see. [3] This network is **essentially** a chain of computers that must all approve an exchange before it can be verified and recorded.

How does it work in practice?

In the case of Bitcoin, block chain stores the details of every transaction of the digital currency, and the technology stops the same Bitcoin being spent more than once.

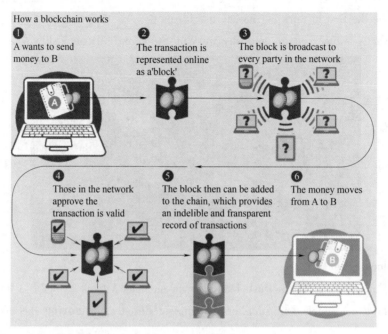

Why is it so **revolutionary**?

The technology can work for almost every type of transaction **involving** value, including money, **goods** and property.④ Its potential uses are almost limitless: from collecting taxes to enabling **migrants** to send money back to family in countries where banking is difficult.

Block chain could also help to reduce **fraud** because every transaction would be recorded and distributed on a public ledger for anyone to see.⑤

 New Words and Expressions

extend [ɪk'stend]　　v. 延伸；延长
currency ['kʌrənsi]　　n. 货币
Silicon Valley　硅谷
remove [rɪ'muːv]　　v. 消除；脱掉
exchange [ɪks'tʃeɪndʒ]　　n. 交换；兑换
decentralized [ˌdiː'sentrəlaɪzd]　　adj. 分散的；去中心化的
essentially [ɪ'senʃəli]　　adv. 本质上；本来的
revolutionary [ˌrevə'luːʃəneri]　　adj. 革命的
involve [ɪn'vɒlv]　　v. 包含；牵涉
goods [gʊdz]　　n. 货物；商品；
migrants ['maɪgrənt]　　n. 移居者；移民
fraud [frɔːd]　　n. 欺骗；诈欺

 Notes

1. Many people know it as the technology behind Bitcoin, but block chain's potential uses

extend far beyond digital currencies.

Analysis："far beyond" 是指 "远超过"。

Translation：很多人都知道它是比特币背后的技术，但区块链的潜在用途远远超出了数字货币。

2. But block chain allows consumers and suppliers to connect directly, removing the need for a third party.

Translation：但区块链允许消费者和供应商直接连接，不需要第三方。

3. Using cryptography to keep exchanges secure, block chain provides a decentralized database, or "digital ledger", of transactions that everyone on the network can see.

Analysis："Using cryptography to keep exchanges secure" 现在分词作伴随状语。

Translation：使用密码学来确保交易的安全，区块链提供了一个去中心化的数据库，或 "数字账本"，网络上的每个人都可以看到交易。

4. The technology can work for almost every type of transaction involving value, including money, goods and property.

Translation：这项技术几乎适用于所有涉及价值的交易，包括货币、商品和财产。

5. Block chain could also help to reduce fraud because every transaction would be recorded and distributed on a public ledger for anyone to see.

Translation：区块链也有助于减少欺诈，因为每一笔交易都会被记录并分发到公共账本上，供任何人查看。

Exercises

Translate the following short passages into Chinese.

1. Many people know it as the technology behind Bitcoin, butblock chain's potential uses extend far beyond digital currencies.

_____

_____

_____

2. Currently, most people use a trusted middleman such as a bank to make a transaction. But block chain allows consumers and suppliers to connect directly, removing the need for a third party.

_____

_____

_____

3. The technology can work for almost every type of transaction involving value, including money, goods and property. Its potential uses are almost limitless: from collecting taxes to enabling migrants to send money back to family in countries where banking is difficult.

Section III　Grammar

Subject Clause and Object Clause
（主语从句和宾语从句）

在复合句中起名词作用的从句叫做名词性从句。它包括主语从句、宾语从句、表语从句和同位语从句。本单元介绍的是主语从句和宾语从句。

引导主语从句和宾语从句的连接词，和名词性从句的连接词一样，均可分为三类：

连接词：that，whether，if（不充当从句的任何成分）

连接代词：what，whatever，who，whoever，whom，whose，which

连接副词：when，where，how，why

一、主语从句

主语从句（Subject Clause）即在主语位置上充当主语的从句。

主语从句通常由从属连词 that，whether，if 和连接代词 what，who，which，whatever，whoever 以及连接副词 how，when，where，why 等词引导。

其中，that 在句中无词义，只起连接作用；连接代词和连接副词在句中既保留自己的疑问含义，又起连接作用，在从句中充当从句的成分。

例如：

What you said is true. （你说的是真的。）

That he was chosen for the job is no surprise. （他被选为这份工作并不令人惊讶。）

Whether we should go or not is up to you. （我们是否应该去取决于你。）

有时为避免句子头重脚轻，常用形式主语 it 来代替主语从句，而把主语从句置于句末。主语从句后的谓语动词一般用单数形式。常用句型如下：

（1）It + be + 名词 + that 从句

（2）It + be + 形容词 + that 从句

（3）It + be + 动词的过去分词 + that 从句

（4）It + 不及物动词 + that 从句

注意：在主语从句中用来表示惊奇、不相信、惋惜、理应如此等语气时，谓语动词要用虚拟语气 "（should）+ do"，常用的句型有：

It is necessary（important，natural，strange，etc.）that…

It is a pity（a shame，no wonder，etc.）that…

It is suggested（requested，proposed，desired，etc.）that…

二、宾语从句

在句中起宾语作用的从句叫宾语从句。

引导宾语从句的关联词与引导主语从句、表语从句的关联词大致一样，在句中可以作谓语动词或介词及非谓语动词的宾语。

1. 由连接词 that 引导的宾语从句。

由连接词 that 引导宾语从句时，that 在句中不担任任何成分，在口语或非正式的文体中常被省去，但如果从句是并列句时，第二个分句前的 that 不可省。

例如：

The study found that regular exercise can reduce the risk of heart disease.

（这项研究发现，经常锻炼可以降低心脏病的风险。）

Many people believe that technology is making us less social.

（许多人认为，技术正在让我们变得不愿意社交。）

注意：在 demand, order, suggest, decide, insist, desire, demand, request, command 等表示 要求、命令、建议、决定等意义的动词后，宾语从句常用"（should）+ 动词原形"。

例如：

The professor insisted that the students（should）submit their essays on time.

（教授坚持要求学生按时提交论文。）

The doctor recommended that I（should）get more exercise and eat a healthier diet.

（医生建议我应该多做运动，吃更健康的饮食。）

2. 用 who, whom, which, whose, what, when, where, why, how, whoever, whatever, whichever 等连接代词引导的宾语从句相当于特殊疑问句，但应注意句子语序要用陈述语序。

例如：

The professor asked who had read the assigned chapter before class.

（教授问谁在上课前已经读了所分配的章节。）

I'm not sure which restaurant we should go to for dinner tonight.

（我不确定今晚我们应该去哪个餐厅吃晚饭。）

The interviewer asked how the candidate would handle a difficult customer.

（面试官问候选人如何处理一个难缠的顾客。）

3. 用 whether 或 if 引导的宾语从句，其主语和谓语的顺序也不能颠倒，仍保持陈述句语序。此外，whether 与 if 在作"是否"的意思讲时，下列情况下一般只能用 whether，不用 if：

a. 引导主语从句并在句首时；

b. 引导表语从句时；

c. 引导从句作介词宾语时；

d. 从句后有"or not"时；

e. 后接动词不定式时。

例如：

The teacher asked whether students had questions about the homework assignment. （老师问学生们是否有关于作业的问题。）

I'm still deciding whether I should apply for the job or not.

（我还在考虑是否应该申请这份工作。）

4. 注意宾语从句中的时态呼应，当主句动词是现在时，从句根据自身的句子情况，而使用不同时态。

例如：I believe that he makes progress every day. （从句用一般现在时）

I believe that he will make progress. （我相信他将会取得进步。）——一般将来时

I believe that he is making progress. （我相信他正在取得进步。）——现在进行时

I believe that he has made progress. （我相信他已取得了进步。）——现在完成时

当主句动词是过去时态（could，would 除外），从句则要用相应的过去时态，如一般过去时、过去进行时、过去将来时等；当从句表示的是客观真理、科学原理、自然现象时，从句仍用现在时态。

例如：The teacher told us that Tom had left us for America.

5. think，believe，imagine，suppose 等动词引起的否定性宾语从句中，要把上述主句中的动词变为否定式。即将从句中的否定形式移到主句中。

例如：I don't think that he has ever been to New York City before.
（我不认为他以前去过纽约市。）

He didn't believe that the movie was worth watching.
（他不相信这部电影值得观看。）

 Exercises

Choose the best answer.

1. _____ makes mistakes must correct them.

   A. What   B. That   C. Whoever   D. Whatever

2. It worried her a bit _____ her hair was turning grey.

   A. while   B. that   C. if   D. for

3. When and why he came here _____ yet.

   A. is not known   B. are not known C. has not known   D. have not known

4. _____ Tom liked to eat was different from _____.

   A. That，that you had expected   B. What，that you had expected

   C. That，what you had expected   D. What，what you had expected

5. _____ we go swimming every day _____ us a lot of good.

   A. If；do   B. That；do   C. If；does   D. That；does

6. He asked me _____.

   A. whether I find out the sender of the money

   B. whether did I find out the sender of the money

   C. whether the sender of the money found out

   D. whether I found out the sender of the money

7. —Do you know _____ the MP3 player last week?

   —Sorry, I have no idea.

A. how much did she pay for   B. how much will she pay for

C. how much she paid for   D. how much she will pay for

8. No one knows _____ the professor will come to our school tomorrow to give us a talk or not.

A. when   B. whether   C. where   D. if

9. Judy didn't know _____ .

A. where is Tim's father   B. when was the first watch made

C. who the old man is   D. what was wrong with her watch

10. He asked me _____ .

A. whether I find out the sender of the money

B. whether did I find out the sender of the money

C. whether the sender of the money found out

D. whether I found out the sender of the money

Section IV　Writing

An E-mail（电子邮件）

　　电子邮件写作是一种在计算机网络上发送和接收信息的书面沟通方式，它与传统的书信写作有所不同。在写电子邮件时，需要注意使用规范的格式、明确的主题、礼貌的称呼、简洁的语言、清晰的表达和适当的结尾等要素，以保证信息的传达和效果.

　　E-mail（Electronic mail）is a method of exchanging messages between people using electronic devices. It is basically seen as the medium of communication to send and receive messages. It is more used for formal purposes than informal. Thus it is the most common form of communication in workplaces around the world. That is why it matters what we write in our e-mail messages and why we need to pay more attention to what our e-mails actually convey.

Part 1　Sample

An E-mail

To：Groupsales@ aston. com

From：LingQiang@ 126. com

Date：November 20, 2018

Reservation Office,

　　I'm writing to book a single room with bath and three standard rooms with bath from Dec. 25 to Dec. 27. And I also want to rent a conference room for business meeting on the afternoon, Dec. 26.

　　Please reply as soon as possible, and let me know if the rooms are available, what the prices are, and if I should pay the deposit. Thank you.

Regards,

Mary

Part 2　Template

Sample One

From：Mary@ ybtk. com. cn

To：Lisa@ ybtk. com. cn

Subject：Study in the U. K.

Dear Lisa,

　　Hi, I am thinking about ＿＿＿＿＿＿＿＿＿（去英国学习）next spring. I have reviewed ＿＿＿＿＿＿＿＿＿＿＿＿（一些著名英国大学的网站）, such as Oxford University and Cambridge University. And I ＿＿＿＿＿＿＿＿＿（已经获得了一些基本资料）.

　　But I need more information about ＿＿＿＿＿＿＿＿＿（会计专业）and ＿＿＿＿＿＿＿＿＿（全额奖学金）. Could you ＿＿＿＿＿＿＿＿＿＿＿（给我一些建议和信息）about my application? And I also want to know _____ (学费和生活费) in Oxford and Cambridge.

    Thank you for your kind help.

Yours,
Mary

Sample Two

From：Will@ sise. com. cn

To：Smith@ sise. com. cn

Subject：Re：Workshop Assistant

Dear Prof. Smith,

    Thanks for the e-mail.

    It is really very kind of you to _____ (邀请我去您夏季的工作室工作). I'm sure I can _____ (得到大量练习) and _____ (提高我的电脑技术). This also gives me a chance to _____ (结交很多新朋友).

    By the way, I'm asked to _____ (教一名小学生数学) on weekend during the summer. I don't know _____ (这是否是一个问题) for your arrangement. Would you please tell me _____ (我应该做什么准备) beforehand?

    Thank you again and look forward to your reply.

Sincerely
Yours,
Will

Part 3　Useful Patterns

1. 我建议我们今晚九点半和布朗先生通一个电话，你和本这个时间可以吗？	1. I suggest that we make a phone call to Mr. Brown at 9：30 pm tonight. Are you and Ben available at this time?
2. 我想跟你在电话里讨论下报告进展和×××项目的情况。	2. I would like to discuss the progress of the report and the ××× project with you over the phone.
3. 我附加了评估报告供您阅读。	3. I have attached an assessment report for your review.
4. 我就文档添加了一些备注，仅供参考。	4. I have added some notes to the document for your reference only.
5. 祝旅途愉快。	5. Have a pleasant journey.
6. 项目进行顺利吗？	6. Is the project going smoothly?
7. 期待您的反馈建议！	7. Looking forward to your feedback and suggestions!

Part 4   Exercises

I. Translate the following e-mail into Chinese.

To：＊＊＊＊＊@＊＊＊＊＊＊.＊＊.＊＊

From：jchen@ ms00. hinet. net

Date：May 3rd，2018

Subject：Thanks for your e-mail

Dear John：

　　Thanks for replying to my e-mail so quickly. It was only 30 minutes ago that I posted my first e-mail, and I already have received your reply. Unbelievable! If it was "snail mail", I would still be on my way back from the post office.

　　You kept telling me about the virtues of going "digital", and now I think I understand what you meant by that. Being able to exchange mail with someone thousands of miles away in 30 minutes or even less is simply amazing.

　　Anyway, I just wanted to say thank you for the e-mail and letting me know that I was doing OK. Like you said, let us keep in touch—online!

Yours,

Jack Chen

II. Write an e-mail according to the information given below.

写信人：Li Hua

写信人地址：LiHua@ hotmail. com

写信时间：2018 年 5 月 22 日

收信人：David

收信人地址：admission-office@ hotmail. com

内容：李华感谢攻读研究生的申请得到接受，但没有能力缴纳学费，希望能获得奖学金帮助自己完成学业。

Unit 10

Workplace and Competition Certificate

Unit Goals

After learning this unit, you will be able to:

- talk about workplace and competition certificate
- understand the passages and grasp the key words and expressions
- review the grammar of subject-verb agreement
- write a farewell letter

Section I　Brief Introduction

　　随着全球化竞争加剧，国内 IT 职场对专业英文能力的要求也在不断提高。对于计算机专业学习者而言，证明自己专业能力的认证——Professional Vocabulary Quotient Credential（PVQC）及 Professional English Listening Comprehension（PELC）成为越来越多学生和职场人士的选择。通过认证考试内容的学习和系统训练，学习者不仅可以熟练掌握计算机专业相关的英语词汇，还能极大提升职场口语交际能力。此外高职学院学生还可以获得相关认证考试证书，优秀的学员还可以参加相关大赛，让学习真正实现"课、赛、证"的融通。

　　PVQC 是由 GLAD 全球学习与测评发展中心，礼聘各专业领域的专家共同指导研发而成。针对在专业工作上的沟通、阅读等问题，将最关键的、经常使用的专业字词做整理，结合技职教育学习理论与现代化云端网络及信息科技，提供各领域的专业人员或大学研究生、技职院校师生有效学习专业英语。

　　PVQC 考试题型与及格标准：

能力与考试类	考试内容	参考等级	通过分数	测验时间	检定方式
阅读和听力 Reading & Listening	测验二：『读』的成绩；看英文，选中文：100 题	总分 Tier One：350～379 分 Tier Two：380～409 分 Tier Three：410～439 分 Tier Four：440～469 分 Tier Five：470～499 分 Tier Six：500 分	总分 350 分以上，且单项成绩不低于 70 分为及格	50 分钟	必考项目
	测验三：『听』的成绩；听英文，选中文：100 题				
	测验四：『听』的成绩；听英文，选英文：100 题				
	测验五：『听』的成绩；看中文，选发音：100 题				
	测验六：『听』的成绩；看英文，选发音：100 题				
拼写 Spelling	测验一：『写』的成绩；看中文，拼写英文：100 题	总分 Tier One：40～49 分 Tier Two：50～59 分 Tier Three：60～69 分 Tier Four：70～79 分 Tier Five：80～99 分 Tier Six：100 分	40	20 分钟	额外选择加考项目

PVQC 分为专业级和专家级，两个级别详细信息如下：

	专业英文词汇专业级	专业英文词汇专家级
专业方向	商业管理、计算机、汽车制造、旅游、餐饮、教育等	
词汇量	约 650 个	约 1 500 个
测试时间	50 分钟 ＋ 20 分钟（选考）	
题目数量	5 个必考科目，每个科目 100 个题目，1 个选考科目，包含 100 题。	
通过标准	每个科目满分都为 100 分，及格成绩为 70 分。 所有 5 个必考科目都通过，才有资格获得国际证书。	
考核形式	在线考试，自动评分	
考核领域	测验 1 看中文，拼写英文（选考） 测验 2 看英文，选中文 测验 3 听英文，选中文 测验 4 听英文，选英文 测验 5 看中文，选发音 测验 6 看英文，选发音	测验 1 看中文，拼写英文（选考） 测验 2 看英文，选中文 测验 3 听英文，选中文 测验 4 听英文，选英文 测验 5 看中文，选发音 测验 6 看英文，选发音
补考规则	在初次考试后 1 个月内，免费补考 1 次	

专业英文对话与听力国际认证（Professional English Listening Comprehension，PELC）是由美国全球学习与测评发展中心（GLAD）推出的世界范围内的专业英文词汇权威国际标准。PELC 国际标准旨在帮助求职者及职场人士快速有效提升专业英文能力，并为自身专业英文水平提供权威证明。PELC 国际标准应用基于大数据分析的技术，遴选行业内最高频率出现的句型，并运用基于最新教育技术的学习系统，对学习者从阅读、听力、发音以及拼写等各个维度进行训练，从而确保学习者能够克服英语学习障碍，在短时间内获得能够切实运用的专业英文能力。

PELC 考试题型与及格标准：

项目	测验名称和说明	题数	时间	满分	及格分	备注
测验 1	句型 1：看图文，选英文	20	5	100	70	必考项目
测验 2	句型 2：听英文，选中文	20	5	100	70	
测验 3	句型 3：看中文，听选英文	20	5	100	70	
测验 4	对话 1：听英文，选英文对话	20	10	100	70	
测验 5	对话 2：看英文，听选英文对话	20	10	100	70	
测验 6	对话 3：听英文，听选英文对话	20	10	100	70	
测验 7	听写：听英文，写英文	10	10	100	40	选考项目
测验 8	译写：看中文，写英文	10	10	100	40	

本项目目前不分级别，详细信息如下：

测试项目	题目数量	测试时间	满分分数	及格分数	备注
测验 1（句型）： 看图文，选英文	20	5 分钟	100	70	所有必考科目都不低于 70 分，才有资格获得国际认证证书
测验 2（句型）： 听英文，选中文	20	5 分钟	100	70	
测验 3（句型）： 看中文，听选英文	20	5 分钟	100	70	
测验 4（对话）： 听英文，选英文对话	20	10 分钟	100	70	
测验 5（对话）： 看英文，听选英文对话	20	10 分钟	100	70	
测验 6（对话）： 听英文，听选英文对话	20	10 分钟	100	70	
测验 7（听写）： 听英文，写英文（选考）	10	10 分钟	100	40	
测验 8（听写）： 看中文，写英文（选考）	10	10 分钟	100	40	
补考规则	在初次考试 1 个月内，免费补考 1 次				

PVQC 与 PELC 的证书样本

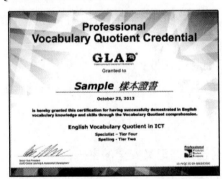

Section Ⅱ Registration and Examination

（Ⅰ）Account Registration Description

考生在参加考试前，必须注册 GLAD 考试账号。下面介绍账号注册的方法。

专业英语系列的在使用专业英语认证系统前，还需要注册一个新的账号，并且激活才能使用。要注册专业英语认证用户账户，请首先打开一个浏览器窗口，在地址栏中输入专业英语认证的主页地址，如下所示。

http：//www.gladcn.com

点击页面下方的"注册"链接，即可进行认证用户的注册（图1）。

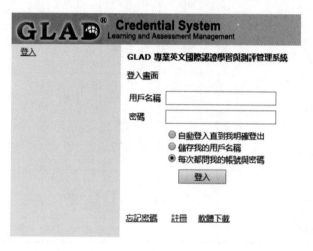

图1

在打开的如图2所示的新用户注册页面中，请参考页面上的注册说明，按规定填写各项内容。

相关身份的编号或学号：此项请直接填写正确身份证号；

英中文姓名：此项目是确定考生在认证书上的名字，请务必认真填写，一旦确定，将不能自行更改；

学校：此项请填写学校名称；

密码/确认密码：此密码为考试登录密码，请选择好记又安全的密码；

E-mail：此项目为考试登录的账号，另外如果忘记密码也需要通过电子邮件来进行找回；

室内电话：此项选填，默认内容即可；

手机：此项选填，默认内容即可；

生日/性别：请根据考生自己的实际情况进行填写即可；

住址：此项选填，默认内容即可；

所属的考试中心或教师：请统一选择"中国北京，CRPH PVQC 管理者3"。

图 2

全部填写完成后，请点击页面下方的"注册"按钮，此时页面将跳转至如下所示的用户注册确认页面，请再次仔细检查所填写项目是否正确。如有任何问题，请点击"取消"按钮后重新注册。如果确认无误，请点击"确认"按钮来完成注册。

当用户注册成功后，系统会返回至登录首页，如图 3 所示，并且在页面上以红色文字显示"注册成功"的标识。此时用户注册即完成。此时，可以使用注册的电子邮箱和密码登录到网站的后台，如图 4 所示。

图 3

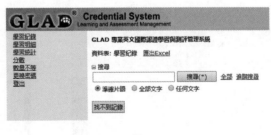
图 4

此时的用户账户，只是注册完成，但并没有激活，不能进行考试。要激活个人的用户账号，请联系报名相关人员来为您处理激活账户的事项。

（Ⅱ）Install Examination and Practice Software

1. 安装专业英语词汇与听力相关考试软件（测评版）

专业英语认证软件下载后均为压缩软件的形式，请将其解压缩到计算机内的某一个位置即可使用。在正式认证考试前，也需要运行一次认证软件，如果有字体不能正常显示或者显示有错误的话，则需要更新系统内的 Unicode 字库，下载并安装 Arial Unicode MS 字体后，系统就会正常运行。

另外需要注意的是，不同的科目和版本，均有不同的软件。请查看后再进行安装和配置。程序安装完毕后，需要向桌面上发送相关软件的快捷方式，PVQC 和 PELC 的考试软件快捷方式如图 5 所示：

专业英文词汇（PVQC）
专业英文听力（PELC）

图 5

2. 安装专业英语词汇与听力相关练习软件（Demo 版）

与测评版安装方法一致，PVQC 和 PELC 的考试软件快捷方式如图 6 所示：

专业英文词汇（PVQC）
专业英文听力（PELC）

图 6

（Ⅲ）Pre Exam Exercise Process

专业英文词汇（PVQC）与专业英文听力（PELC）在练习方式上有所不同，下面做详细介绍。

（一）专业英文词汇（PVQC）考前练习

1. 使用 PVQC Demo 版进行练习

需要注意的是，不同的科目和版本，均有不同的软件，请打开对应科目的 Demo 软件后，直接点击 Login 进入系统（图 7）。图 8 为练习界面，选择需要练习的项目，点击 OK 即可进入练习。

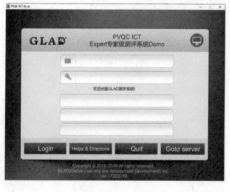

图 7
图 8

2. 使用 PVQC 测评版（考试软件）进行练习

需要注意的是，不同的科目和版本，均有不同的软件，请打开对应科目的测评版软件后，输入已注册的账号和密码并激活成功后，点击 Login 进入系统，并选择想要练习科目（图9）。图10为练习界面，选择需要练习的项目，点击 OK 即可进入练习。

图9　　　　　　　　　　　　　　　　　　　图10

3. PVQC 测评版（考试软件）和 Demo 版（练习软件）的区别

版本	功能划分	能否进行考试	优　点	缺　点
PVQC Demo 版	只能练习	不能进行考试	无须注册账号即可免费试用	练习模式每个单项每次出题20道，需要耗费大量时间刷题
PVQC 测评版	可以练习，也能考试	可以进行考试	注册账号并激活后，可以进行完整练习（练习模式每个单项每次出题100到）提高刷题速度	需要提交账号注册信息，缴费后并激活才能使用

（二）专业英文听力（PELC）进行考前练习

使用 PELC Demo 版进行练习

PELC 只有一个科目，使用前请打开 PELC Demo 版软件，输入已注册的账号和密码并激活成功后，点击 Login 进入练习系统（图11）。

图11

(Ⅳ) Examination Process

考试形式为在线上机考试，请考生确认网络畅通，并在 Windows 系统下运行考试程序。

在考生参加比赛考试之前，须首先注册考试账号，注册流程请查看上文介绍的账号注册相关内容。

在考生就座后，双击桌面上的 PVQC 或 PELC 考试的快捷图标即可进入考试登录画面，然后按照监考老师的指示，输入选手注册的用户名和密码，然后继续输入认证码（需提前申请报备）、监考人用户和选手身份证号后五位。最后点击登录按钮。图 12 是 PVQC 认证系统，图 13 是 PELC 认证系统。

图 12　　　　　　　　　　　　　图 13

PELC 和 PVQC 考试流程一致，下面以 PVQC 为例，介绍考试流程。

登录后，选择要参加考试的科目，PVQC 词汇考试，考试项目是科目二到科目六，科目一为选考内容（图 14）；PELC 听力考试，考试项目是科目一至科目六，科目七到科目八为选考内容（图 15）。

图 14　　　　　　　　　　　　　图 15

选择完考试科目后，点击开始测试，就开始了正式考试的环节。在这里请根据题目要求来进行选择，在选择的时候，可以使用鼠标点击，也可以使用数字键盘对应的数字进行选择。除了 PVQC 的测验一和 PELC 的测验七、测验八以外，都可以使用鼠标点击耳机图标来播放对应的读音或语句（图 16、图 17）。

图 16

图 17

每当作答完一个测验后，会有该测验的分数，以及正确和错误的题目数量。查看后，点击【下一测验】，来继续答题（图18、图19）。

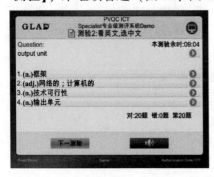

图 18

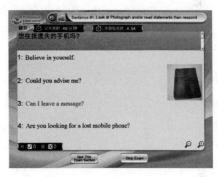

图 19

全部作答完毕后，会显示成绩单。成绩单上会显示单项成绩、作答时间，以及总成绩和总用时。同时还可以查看做错的题目的正确答案（图20、图21）。

图 20

图 21

这时候，请查看成绩下方是否提示"您的成绩已成功上传服务器"，如果有这个提示，说明考试成功结束，否则请联系监考老师处理。考试结束后，请点击【返回】按钮回到主菜单，最后点击【离开】按钮，按照屏幕提示退出考试系统。

最后，请登录 www.gladcn.com 网站，查看个人成绩是否已经成功上传，确认无误后，即可关闭桌面所有窗口后，完成考试。

Section III Simulation Test

PVQC Specialist（专业级）模拟测验（一）
核心词汇

选填：把与英文相对应的中文释义的字母代码写在横线上。

A. 信息与通信技术	B. 硬件	C. 处理器	D. 字节
E. 服务器	F. 芯片	G. 计算机	H. 系统
I. 升级	J. 因特网	K. 软件	L. 数字
M. 装置	N. 文字	O. 内存	P. 传感器
Q. 打印	R. 高速缓存	S. 动力；电源	T. 软件
U. 端口	V. 应用软件	W. 数据	X. 工作站
Y. 键盘			

1. hardware _____
2. processor _____
3. byte _____
4. computer _____
5. Internet _____
6. chip _____
7. upgrade _____
8. server _____
9. ICT _____
10. system _____
11. software _____
12. digital _____
13. memory _____
14. device _____
15. software _____
16. cache _____
17. power _____
18. sensor _____
19. print _____
20. port _____
21. data _____
22. application _____
23. text _____
24. workstation _____
25. keyboard _____

PVQC Expert（专家级）模拟测验（二）
词汇练习

在下面的横线上填写相关的英文或中文。

1. 数据管理系统缩写为_____

2. data type _____

3. 数据域目；数据字段_____

4. 数据摘要_____

5. data model _____

6. 数据词典，缩写为 DD _____

7. program – data independence _____

8. a conceptual representation _____

9. 数据库管理师，缩写为 DBA _____

10. end users _____

11. system analyst _____

12. 软件工程师_____

13. 可携式文件格式，缩写为 PDF _____

14. 可携型计算机；笔记式计算机_____

15. 可视化程序（设计）_____

16. virus protection and detection _____

17. 文件传输协议，缩写为 FTP _____

18. 传输协议，缩写为 TCP _____

19. 互联网通信协议_____

20. 服务质量，缩写为 QoS _____

PELC 专业英语能力测验模拟考题（一）
范围：工作申请与面试

Job Application and Interviews（JAI）

班级：　　　姓名：　　　学号：

英文句翻母语（native language）练习（每题 5 分，共 100 分，70 分（含）以上及格）

（　　）1. Good luck on the interview.

A. 祝你期中考试顺利。

B. 祝你好运。

C. 祝面谈（口试）顺利。

D. 很高兴认识你。

（　　）2. Please go to the room in the corner and wait there.

A. 请到二楼并且待在那边。

B. 请到转角的房间等一下。

C. 请客会花我许多等待的时间。

D. 请不要客气。

（　　）3. Do you have any licenses or certificates?

A. 您审阅过他的履历表了吗？

B. 您觉得需要花多久时间？

C. 您有哪些执照或是证照？

D. 您有哪些职位或是保证？

（　　）4. Tell me about yourself.

A. 告诉我关于您自己。

B. 请到转角的房间等一下。

C. 我告诉自己要很小心。

D. 请不要客气。

（　　）5. Did you review her resume?

A. 您审阅过她的履历表了吗？

B. 您觉得需要花多久时间？

C. 您有哪些执照或是证照？

D. 您有哪些职位或是保证？

（　　）6. Nice to meet you.

A. 告诉我关于您自己。

B. 多爱护你自己。

C. 告诉自己要很小心。

D. 很高兴认识你。

（　　）7. How long do you think it will take you?

A. 您审阅她的履历表花了多久时间？

B. 您觉得需要花多久时间？

C. 您有哪些执照或是证照？

D. 您有哪些职位或是保证？

（　　）8. How long will it take me to get to the airport?

A. 您审阅她的履历表花了多久时间？

B. 到机场会花我多少时间？

C. 从机场到车站有多长？

D. 到机场的距离有多少？

（　　）9. How does she handle a conflict situation?

A. 她有不高兴的情形吗？

B. 她如何控制场面？

C. 到机场会花她多少费用？

D. 她如何处理冲突的情况？

（　　）10. Is there anything you want me to do before starting?

A. 您一开始有任何不高兴的事情吗？

B. 在开始前，您有任何事要我先做吗？

C. 在完成后，您要我转达什么意见吗？

D. 您是如何处理冲突的情况？

（　　）11. I believe I can use my education and experience in this job.

A. 在工作中，我信任我的家人与同事的经验。

B. 我相信我的朋友能够接纳我的工作。

C. 我相信专业工作态度可适用于所有工作职位。

D. 我相信我能够运用我受的教育与经验在这个工作中。

（　　）12. I think a professional attitude towards work applies to all job positions.

A. 对所有工作而言，我信任我同事的能力。

B. 我认为专业工作态度可适用于所有工作职位。

C. 我相信专业工作态度让我迈向成功之路。

D. 我相信我能够运用我受的教育与经验在所有工作中。

（　　）13. I think these skills apply to all job positions.

A. 对所有工作而言，我信任我同事的能力。

B. 我认为专业工作态度可适用于所有工作职位。

C. 我认为这些技能适用于所有的工作职位上。

D. 我相信我能够运用我受的教育与经验在所有工作中。

（　　）14. I will prove to you that I can perform very well.

A. 我会证明给您看我能表现良好。

B. 我认为专业工作态度可适用于所有工作职位。

C. 我认为这些技能适用于所有的工作职位上。

D. 我相信我能够运用我的良好态度与经验在所有工作中。

（   ）15. I would like my job to be more challenging.

A. 我会证明给您看我能表现良好。

B. 我认为专业工作态度可适用于所有工作职位。

C. 我认为这些技能适用于所有的工作职位上。

D. 我喜欢我的工作更具挑战性。

（   ）16. I'm signing up for a few multimedia design and maker classes.

A. 我正在报名一些手机及语言方面的课程。

B. 我认为专业工作态度可适用于所有工作职位。

C. 我正在报名一些多媒体及创客（自造者）方面的课程。

D. 我喜欢我的工作更具挑战性。

（   ）17. I am highly motivated and well organized.

A. 我正在报名一些手机及语言方面的课程。

B. 我认为专业工作态度可适用于所有工作职位。

C. 我正在报名一些多媒体及创客方面的课程。

D. 我机动性强且具有良好的组织能力。

（   ）18. My major is App design.

A. 我主修手机应用软件设计。

B. 我主要的工作是一些手机的设计。

C. 我正在报名一些多媒体及创客（自造者）方面的课程。

D. 我机动性强且具有良好的组织能力。

（   ）19. He pays a lot of attention to details.

A. 他主修手机应用软件设计。

B. 他主要的工作是一些手机的设计。

C. 他很重视细节。

D. 他花了很多心血。

（   ）20. These experiences familiarize me with the market in Asia.

A. 我的家人很喜欢亚洲的市场。

B. 这些经验让我熟悉亚洲的市场。

C. 这些款式很受亚洲市场欢迎。

D. 我花了很多心血在亚洲的主题研究。

PELC 专业英语能力测验模拟考题（二）
范围：工作申请与面试

Job Application and Interviews（JAI）

班级：　　　姓名：　　　学号：

母语（native language）中文句翻成英文句（每题 5 分，共 100 分，70 分（含）及格）

（　　）1. 面谈（口试）顺利。

A. Good luck on the midterm exam.

B. Good luck to you.

C. Good luck on the interview.

D. Nice to meet you.

（　　）2. 您能为本公司提供什么？

A. What can you offer to the company?

B. What do you want me to do?

C. What is this woman's occupation?

D. What will you do first?

（　　）3. 您将先做什么？

A. What can you offer to the company?

B. What do you want me to do?

C. What is this woman's occupation?

D. What will you do first?

（　　）4. 我应该先做什么？

A. What should I do first ?

B. What do you want me to do?

C. What is this woman's occupation?

D. What will you do first?

（　　）5. 您有（开放）任何工作职位吗？

A. What should I do first ?

B. Do you have any job openings?

C. What is this woman's occupation?

D. What will you do first?

（　　）6. 这位女士的职业（务）是什么？

A. What should I do first ?

B. Do you have any job openings?

C. What is this woman's occupation?

D. What will you do first?

（　　）7. 您对本公司有什么认识？

A. What should I do first ?

B. Do you have any job openings?

C. What is this woman's occupation?

D. What do you know about this company?

(   ) 8. 您对薪水有什么要求？

A. What should I do first ?

B. What are your salary expectations?

C. What is this woman's occupation?

D. What do you know about this company?

(   ) 9. 她是做什么样工作的？

A. What kind of work does she do?

B. What are your salary expectations?

C. What is this woman's occupation?

D. What do you know about this company?

(   ) 10. 您提供了什么职位？

A. What kind of work does she do?

B. What are your salary expectations?

C. What positions are you offering?

D. What do you know about this company?

(   ) 11. 您有工作经验吗？

A. What kind of work does she do?

B. What are your salary expectations?

C. When can she embark on a new career?

D. Do you have any work experience?

(   ) 12. 她何时可着手（开展）新的（职业）生涯？

A. What kind of work does she do?

B. What are your salary expectations?

C. When can she embark on a new career?

D. Do you have any job openings?

(   ) 13. 为什么这位男士很在意？

A. Why is this man concerned?

B. What are your salary expectations?

C. When can she embark on a new career?

D. Do you have any job openings?

(   ) 14. 您为什么有兴趣来我们公司工作？

A. Why is this man concerned?

B. Why are you interested in working for our company?

C. When can she embark on a new career?

D. Do you have any job openings?

(　　) 15. 您为什么离开上一份工作？

A. Why is this man concerned?

B. Why are you interested in working for our company?

C. Why did you leave your last job?

D. When can she embark on a new career?

(　　) 16. 您为什么想来这家公司工作？

A. Why is this man concerned?

B. Why are you interested in doing this for third party?

C. Why did you leave your last job?

D. Why do you want to work for this company?

(　　) 17. 您当时为何做此决定？

A. Why did you make the decision then?

B. Why are you interested in doing this for third party?

C. Why did you leave your last job?

D. Why do you want to work for this company?

(　　) 18. 我们何不下午去看工厂？

A. Why did you make the decision then?

B. Why are you interested in working for our company?

C. Why don't we go to the factory this afternoon?

D. Why do you want to work for this company?

(　　) 19. 我们为什么要雇用你？

A. Why should we hire you?

B. Why are you interested in working for our company?

C. Why don't we go to the factory this afternoon?

D. Why do you want to work for this company?

(　　) 20. 在这件商务上，为什么我们要相信你？

A. Why did you make the decision then?

B. Why are you interested in working for our company?

C. Why don't we go to the factory this afternoon?

D. Why should we believe you on this business?

Section IV Grammar

Swbject-Verb Agreement（主谓一致）

主谓一致就是谓语动词必须在数和人称上与主语取得一致。主谓一致的关系根据"语法一致""意义一致"和"就近一致"三项原则来实现。

一、语法一致

◆ 主语和谓语从语法形式上取得一致：主语是单数形式，谓语也采取单数形式；主语是复数形式，谓语亦采取复数形式。例如：

A grammar book helps you learn something about the rules of a language.

（主语是单数形式，谓语也采取单数形式）

语法书帮助你学习语言的某些规则。

Grammar books help you learn something about the rules of a language.

（主语是复数形式，谓语也采取复数形式）

语法书帮助你学习语言的某些规则。

◆ 主语和谓语从语法形式上取得一致的问题远不止上述的那么简单，有许多方面的情况需要去具体地对待：

1. 不定式、动名词，以及从句作主语时应看作单数，谓语动词用单数。例如：

Reading often means learning.

读书常意味着学习。

To read English aloud every morning does you a lot of good.

每天早晨高声朗读英语有许多好处。

What he said has been recorded.

他说的话已被录音了。

2. 不定代词 one, every, each, everybody, everyone, one of, no one, nothing, nobody, someone, somebody, either, neither, many a 等作主语或是修饰主语时应看作单数，谓语动词用单数。例如：

Neither of my sisters likes sports.

我的两个妹妹没人喜欢运动。

Many a student takes a walk on campus after dinner.

许多学生晚饭后常在校园里散步。

Every boy and girl shows great interest in extra-curriculum activities.

每个男孩和女孩对课外活动都表现出很大的兴趣。

3. 表示国家、机构、事件、作品等名称的专有名词作主语时应看作单数，谓语动词用单数。例如：

One Thousand And One Nights tells people lots of mysterious bits of folklore.

《一千零一夜》给人们还讲述了许多神秘的民间传说。

The United States is leading the world in science and technology.

美国在世界科技方面领先。

The United Nations plays an important role in the international affairs.

联合国在国际事务中起着重要作用。

4. a portion, a series of, a kind of, the number of 等与名词构成名词短语作主语时应看作单数，谓语动词用单数。例如：

A series of high technology products has been laid out in the exhibition.

一系列高科技产品已在展览上展出。

The number of printing mistakes in some recent books often surprises people even to death.

近来一些书籍里印刷错误的数量让人吃惊得要命。

A substantial portion of the reports is missing.

这些报告都没有提及实质问题。

A kind of rose in the garden smells very pleasant.

这座花园里有一种玫瑰香气怡人。

5. 由 some, several, both, few, many, a number of 等词修饰主语，或是由它们自身作主语时应看作复数，谓语动词用复数。另外，由 and 连接两个主语时，谓语一般用复数。例如：

On the seashore, some people are playing volleyball and some are lying in the sun.

海边，有些人在打排球，有些人躺着晒太阳。

Both of us are fond of watching football games.

我们俩都喜欢看足球赛。

A number of will-be graduates are voluntarily going to work in the West of China.

许多即将毕业的学生打算自愿去中国西部工作。

6. 有些短语，如：a lot of, most of, any of, half of, three fifths of, eighty percent of, some of, none of, the rest of, all of 等后接不可数名词或是单数形式的名词作主语时应看作单数，谓语动词用单数；但如果后接可数名词的复数形式作主语时应看作复数，谓语动词用复数。例如：

A lot of money in the shop was stolen yesterday when the electricity was suddenly cut off.

昨天突然断电时，那家商店丢失了许多钱。

A lot of books about Investment Fund have been published recently.

最近出版了许多关于投资基金的书籍。

二、意义一致

这一原则是指，从意义着眼来解决主谓一致问题。有时主语形式上为单数，但意义上却是复数，那么谓语依意义也用复数形式；而有时主语形式上为复数，但意义上却是单数，那么谓语依意义亦用单数形式。

1. 当主语后面接由 as well as, as much as, accompanied by, including, in addition to, more than, no less than, rather than, together with 等引导的词组时，其谓语动词的形式要依主语的

单复数而定。在这样的句子里，这些词所引导的词组不影响主语自身的单、复数形式，它们在句子里其实是状语。也就是说，我们完全可以将这些词组搬到句首或是放到句末去。从表面上我们也可以看出，它们与主语之间有"，"隔开。例如：

Petroleum, along with fuel gas, has recently risen in price.

最近石油和燃料煤气的价格上涨了。

The teacher, with all his students, is going to have a picnic this weekend.

老师打算这个周末与学生们一起去野炊。

The students, together with their teacher, are going to have a picnic this weekend.

学生们打算这个周末与他们的老师一起去野炊。

The warehouse, with all its stockings, was burned last night.

昨晚，那个仓库连同其所有的货物一起被烧毁了。

2. 表示时间、金钱、距离、体积、重量、面积、数字等词语作主语时，其意义若是指总量应看作单数，谓语动词用单数；但如果其意义是指"有多少数量"则应该看作是复数，那么谓语动词也应该用复数。例如：

Four weeks are often approximately regarded as one month.

人们通常大约地将四个星期看成一个月。

Twenty years stands for a long period in one's life.

二十年在人的一生里意味着一个很长的时期。

Eighty dollars are enough for a student to spend on food for one week.

八十块钱足够让一个学生吃一个月的饭了。

3. 形容词前加定冠词即"the + 形容词"作主语时，其意义若是指个人或是抽象概念应看作单数，谓语动词用单数；但如果其意义是指一类人则应该看作是复数，那么谓语动词也应该用复数。例如：

The young, on one hand, often think of the old conservative. On the other hand, the old always consider the young inexperience.

一方面，青年人常认为老年人保守；另一方面，老年人总是认为青年人没有经验。

In many stories, the good are well rewarded and the bad are doomed to unfortunate.

在许多故事里，好人总是有好报，坏人注定要倒霉。

4. 当 and 连接两个并列主语在意义上指同一人、同一物、同一事或者同一概念时，应看作单数，谓语动词用单数。另外，当 and 连接两个形容词去修饰一个单数形式的主语时，其实是指两种不同的事物，主语则应该看作是复数，那么谓语动词也应该用复数。例如：

War and peace is a constant theme in history.

战争与和平是一个历史上的永恒的主题。

Chinese and Japanese silk are of good quality.

中国丝绸和日本丝绸质量都很好。

Different people respectively welcome white and black coffee.

加奶的咖啡与清咖啡都分别受到不同人们的喜爱。

5. 集体名词作主语时，谓语动词的数取决于主语的意义：主语表示整体时视为单数，谓语动词用单数；主语表示集体中的个体成员时视为复数，谓语动词用复数。这类集体名词常见的有：army, audience, cattle, class, club, committee, crowd, family, government, group, majority, minority, part, people, police, public, staff, team 等等，其中 cattle, people, police 一般看成复数形式。例如：

The family are all fond of football.

那一家人都喜欢足球。

The family is the tiniest cell of the society.

家庭是社会的最小的细胞。

The public has every reason to be cautious of professional deception.

人民大众完全有理由谨防职业骗局。

The public now come to know the whole story.

人们现在越来越清楚那是怎么回事了。

三、就近一致

这一原则是指，谓语动词的人称和数常常与最近作主语的词语保持一致。常出现在这类句子中的连词有：or, either...or..., neither...nor..., not only...but also 等。例如：

Either I or they are responsible for the result of the matter.

不是我，就是他们要对那件事的结果负责。

Neither the unkind words nor the unfriendly attitude has caused me any distress.

既不是那些不友好的话，也不是那不友好的态度让我沮丧。

Not only he but also all his family are keen on concerts.

不仅仅是他，他全家人都很热衷于音乐会。

Neither his family nor he knows anything about it.

他家人和他都不知道那件事。

 Exercises

Choose the best answer.

1. One-third of the area _____ covered with green trees. About seventy percent of the trees _____ been planted.

    A. are; have   B. is; has   C. is; have   D. are; has

2. The number of teachers in our college _____ greatly increased last term. A number of teachers in this school _____ from the countryside.

    A. was; is   B. was; are   C. were; are   D. were; is

3. What _____ the population of China? One-third of the population _____ workers here.

    A. is; are   B. are; are   C. is; is   D. are; is

4. Not only he but also we _____ right. He as well as we _____ right.

    A. are；are   B. are；is   C. is；is   D. is；are

5. What he'd like _____ a digital watch. What she'd like _____ textbooks.

    A. are；are   B. is；is   C. is；are   D. are；is

6. He is one of the boys who _____ here on time. He is the only one of the boys who _____ here on time.

    A. has come；have come   B. have come；has come

    C. has come；has come   D. have come；have come

7. Either you or he _____ interested in playing chess. _____ you or he fond of music at present?

    A. are；Are   B. is；Are   C. are；Is   D. is；Is

8. Many a professor _____ looking forward to visiting Germany now. Many scientists _____ studied animals and plants in the last two years.

    A. is；have   B. is；has   C. are；have   D. is；are

9. A knife and a fork _____ on the table. A knife and fork _____ on the table.

    A. is；is   B. are；are   C. are；is   D. is；are

10. My friend and classmate Paul _____ motorcycles in his spare time.

    A. race   B. races   C. is raced   D. is racing

Section V　Writing

辞职信

辞职信是辞职者向工作单位辞去职务时写的书信，也叫辞职书或辞呈。辞职信是辞职者在辞去职务时的一个必要程序，通常由标题、称谓、正文、结语、署名与日期五部分构成。

Part 1　Sample

A Farewell Letter

Dear Mr. Wong,

　　I would like to let you know how much I have enjoyed my last three years at Hero Company. Hero Company is an invaluable place for enriching my knowledge about financial field, I enjoyed working with my colleagues and I have learned so much things here.

　　Since I would like to take a new challenge and I want to meet people from all walks of life, I have accepted an offer from an Insurance firm as a Personal Financial Consultant. I would therefore appreciate it if you would accept my resignation effective from 8 March, 2002.

　　I would be very much obliged if you would kindly give me a reference letter before I leave. Thank you for all that you have done to make my work here both interesting and enjoyable.

Yours sincerely, Alexander Fung

Part 2　Template

Sample One

Dear _____,

　　I am writing to inform you about my decision of resignation. I have enjoyed working with you and the staff in ourcompany in the past few years. However, I find it is inappropriate for me to take the position as _____（接职务，如 an editor）for the following reasons.

　　First and foremost, _____（第一个原因）. Besides, _____（第二个原因）. Most importantly, _____（第三个原因）.

　　Thank you for your caring in these several years working. I am deeply sorry for any inconvenience my leaving may cause.

Yours sincerely,

Li Ming

Sample Two

Dear Mary,

　　I would like to take this opportunity to _____（和你们所有人道别）. I am _____（处于离职程序中）at Goldman's. This coming Friday is my last day in attendance.

I have _____ （相当享受在这个公司工作）and appreciate your friendship during these many years. _____ （感谢你的支持和鼓励），without which I'd never have been able to reach the goals I've set for myself. While I look forward to this new position, and the challenges it entails, I will truly miss those of you here.

I _____ （希望你和公司在将来取得每一个成功）. Please feel free to contact me any time. I'll also like to add you to my professional, online profile at LinkedIn. These are _____ （我的联系方式）: sdr@ None. com；XdG@ LinkedIn. com; 555-234-0033.

Best Regards,

Benedict

Part 3 | Useful Patterns

1. 我想跟你告别并告诉你我将离开我在…的职位。

2. 我下周就要离开这家公司去寻找其他的职业机会了。

3. 正如我在上周的会议上告诉你的，我在……的最后一个工作日是下周五。

4. 在过去的两年里，您给了我支持和帮助。

5. 在您的鼓励和指导下，我做到了出色完成提供给我的项目。

6. 我很感激曾拥有这个和大家一起工作的好机会。

7. 感谢您这两年的支持和鼓励。

1. I want to bid farewell to you and inform you that I am leaving my position at…

2. I am bidding adieu next week to this company to pursue other career opportunities.

3. As I informed you in our meeting last week, my last working day at…is next Friday.

4. During these last two years you have provided me support and favor.

5. Through your encouragement and guidance I have been able to excel at the projects offered to me.

6. I appreciate having had this wonderful opportunity to work with you all.

7. I thank you for your support and encouragement during these two years.

I . Translate the following part of a farewell letter into Chinese.

Dear Tax,

The time have finally arrived, when I must bid Adieu to all of my most beloved colleagues and friends.

After spending these many years together, it will be difficult to part from you, and this company that I've come to know so well. I've found our association to be productive as well as encouraging. I hope to take all that I've learned here and apply it to my retirement years.

I wish you and yours continued success on the roads you continue to travel. My best to the managerial staff and the company as well. This is not a "Goodbye", but simple a pause in the road. Thank you.

Please stay in touch: e-mail: × × ×@None. com; phone: 555-555-2345. Good luck to each and every one of you,

Best Regards,

Jay

II . Write a farewell letter according to the information given below.

假设你是 ABC 公司的 Samantha，你即将离开这个公司，并且下个月将在 XYZ 公司开始一个新职位。请你写一封告别信给同事 John，告诉他你很享受之前在这里的工作，也很感谢有机会与之共事，更感谢他在你任职 ABC 公司期间给予你的支持、指导和鼓励。你还可以在信中用其他细节表现出你对这份工作的想念和对新工作的展望。最后，不要忘记留下你的联系方式以便他将来和你联系，你的邮箱是：samantha83@ gmail2. com，你的电话号码是 555-555-2222。

参考词汇：任期 tenure；联系 reach